Language, Literacy, and Learning in STEM Education

Language Studies, Science and Engineering (LSSE)

The *Language Studies, Science and Engineering* (LSSE) book series seeks submissions of book proposals that address the interface among language studies, science, engineering and education. This book series aims to bring together researchers from the fields of language studies and science/engineering education with the aim of generating new interdisciplinary knowledge. This book series is premised on the concept that science is of central importance in the 21st century and that research informed by linguistic knowledge can contribute to the description, understanding, education and practice of science and engineering. The goal of this series is to enhance educational and professional practices in the sciences and engineering through interdisciplinary interaction between language researchers, science and engineering educators and scientists.
The scope of this book series covers the range of potential contributions that language studies can make to the advancement of science, engineering and educational practices in these fields. Researchers who utilize language based methodologies, such as discourse analysis, computational linguistics, conversational analysis, multimodal analysis, rhetorical analysis, and genre analysis, for the purposes of advancing science/engineering education and professional practice are invited to submit a proposal for this innovative book series.

For an overview of all books published in this series, please see
http://benjamins.com/catalog/lsse

Volume 1

Language, Literacy, and Learning in STEM Education. Research Methods and Perspectives from Applied Linguistics
Edited by Mary Jane Curry and David I. Hanauer

Language, Literacy, and Learning in STEM Education

Research Methods and Perspectives from Applied Linguistics

Edited by

Mary Jane Curry
University of Rochester

David I. Hanauer
Indiana University of Pennsylvania

John Benjamins Publishing Company
Amsterdam / Philadelphia

™ The paper used in this publication meets the minimum requirements of the American National Standard for Information Sciences – Permanence of Paper for Printed Library Materials, ANSI z39.48-1984.

Library of Congress Cataloging-in-Publication Data

Language, Literacy, and Learning in STEM Education : Research Methods and Perspectives from Applied Linguistics / Edited by Mary Jane Curry and David I. Hanauer.

p. cm. (Language Studies, Science and Engineering, ISSN 2210-7029 ; v. 1)

Includes bibliographical references and index.

1. Applied linguistics. 2. Language acquisition. 3. Interdisciplinary approach in education. 4. Science--Study and teaching. 5. Technology--Study and teaching. 6. Engineering--Study and teaching. 7. Mathematics--Study and teaching. I. Curry, Mary Jane, editor of compilation. II. Hanauer, David Ian, editor of compilation.

P129.L367 2014

418.0071--dc23 2014011080

ISBN 978 90 272 5750 5 (Hb ; alk. paper)

ISBN 978 90 272 7011 5 (Eb)

John Benjamins Publishing Co. · P.O. Box 36224 · 1020 ME Amsterdam · The Netherlands

John Benjamins North America · P.O. Box 27519 · Philadelphia PA 19118-0519 · USA

Table of contents

List of figures

List of tables

Contributor biographies

Cynthia Bauerle is Assistant Director in Undergraduate and Graduate Science Education at the Howard Hughes Medical Institute, where she manages the institute's science education portfolio of grants, fellowships, and special initiatives. She coordinates the NEXUS project and other multi-institutional initiatives focused on improving undergraduate science education and student persistence in STEM disciplines. Bauerle spent two decades in higher education as a faculty member primarily at undergraduate institutions and was a Fulbright Senior Scholar at the University of Dar es Salaam, Tanzania, in 1999–2000. She earned her undergraduate degree from the University of Virginia and her Ph.D. in molecular biology from the University of Wisconsin-Madison.

Jeffrey Choppin directs the mathematics education program at the Warner Graduate School of Education at the University of Rochester. Choppin's research currently focuses on teachers' perceptions and uses of curriculum materials in the context of the Common Core State Standards for Mathematics (CCSSM). His research also considers what teachers learn from using innovative curriculum materials, particularly with respect to teacher knowledge of how the materials might develop student reasoning around key mathematical concepts. His research links teachers' discourse practices, observations of student thinking, understanding of the curriculum materials, and the ways their adaptations of curriculum materials enhance students' opportunities to engage in mathematical practices.

Mary Jane Curry is Associate Professor at the Warner School of Education, University of Rochester. She is co-author of *A scholar's guide to getting published in English: Critical choices and practical strategies* (Multilingual Matters, 2013), *Academic writing in a global context: The politics and practices of publishing in English* (Routledge 2010), and *Teaching academic writing: A toolkit for higher education* (Routledge 2002). She has published in *Written Communication, TESOL Quarterly, English for Specific Purposes,* and *the Journal of English for Academic Purposes.* She is principal investigator for a 2012–2017 National Professional Development grant from the U.S. Department of Education. In 2014 she was a Fulbright Fellow in Chile.

Julio Gimenez is a Lecturer at the University of Nottingham, United Kingdom. His main research interests are in academic literacies with a special focus on disciplines such as business, health sciences, and engineering, and the role of language in academic and professional contexts. His work has appeared in international journals and edited collections in areas of his research, including the *Journal of English for Academic Purposes, Higher Education, Journal of Learning Development in Higher Education* and the *European Journal of Engineering Education.*

David I. Hanauer is Professor of Applied Linguistics/English at Indiana University of Pennsylvania and Assessment Coordinator of the Phage Hunters Integrating Research and Education Program at the University of Pittsburgh. He is the author of six books, including *Scientific discourse: Multiliteracy in the classroom, Scientific writing in a second language* and *Active assessment: Assessing scientific inquiry.* He has published in *Science, CBE-LSE* and a wide range of applied linguistics and education journals. Dr. Hanauer has received funding from the National Science Foundation, the Howard Hughes Medical Institute, and the United States Department of Education. He is currently the series editor of Language Studies, Science and Engineering for John Benjamins Publishers and editor of *Scientific Studies of Literature.*

Graham Hatfull is Professor of Biological Sciences and Eberly Family Professor of Biotechnology at the University of Pittsburgh. He received his Ph.D. from the University of Edinburgh and did postdoctoral research at the MRC labs in Cambridge, UK, and at Yale University, before joining the University of Pittsburgh as Assistant Professor in 1988. Dr. Hatfull has been a Howard Hughes Medical Institute Professor since 2002. His research interests focus on the molecular genetics of the mycobacteria and their bacteriophages and the development of phage discovery and genomics as a broadly implementable platform for science education.

Joana Latas has an undergraduate degree in mathematics and computer science from the University of Evora, Portugal, and a master's in educational sciences from the same university, where she did research in ethnomathematics. She collaborates with the Centre for Research in Education and Psychology of the University of Evora (CIEP-UÉ). Since 2013 she has been the representative for Mathematics of Planet Earth in the Republic of São Tome and Príncipe, a country where she has taught mathematics in high school and mathematics education in higher education. She is currently a Ph.D. student at University of Coimbra, Portugal, in history of science and scientific education.

Neal Lerner is Associate Professor of English and Director of Writing in the Disciplines at Northeastern University. His book, *The idea of a writing laboratory*, won the 2011 National Council of Teachers of English David H. Russell Award for Distinguished Research in the Teaching of English. He is also co-author of *Learning to communicate as a scientist and engineer: Case studies from MIT*, winner of the 2012 Conference on College Composition and Communication Advancement of Knowledge Award, and *The Longman guide to peer tutoring*. He has published on the history, theory, administration, and practice of teaching writing in classrooms, laboratories, and writing centers.

Darlinda Moreira earned her doctorate at ISCTE (Instituto Universitário de Lisboa) in anthropology of education, her master's degree in Bilingual Education Studies (University of Massachusetts at Boston) and her undergraduate degree in mathematics (University of Lisbon). Presently she is Professor and the chair of the Department of Education and Distance Education at the Universidade Aberta, Portugal (The Open University). Her research interests lie in the area of education in global contexts and interculturality, lifelong learning, and cyberculture and education. She has extensive experience in tertiary education and teacher training and is the author of many articles and book chapters.

Mya Poe is Assistant Professor of English at Northeastern University. Her research focuses on writing assessment, diversity, and writing in the disciplines. She is co-author of *Learning to communicate in science and engineering: Case studies from MIT*, which won the Conference on College Composition and Communication 2012 Advancement of Knowledge Award, co-editor of *Race and writing assessment*, and series co-editor of *The Oxford short guides to writing in the disciplines*. She is currently working on a book entitled "The consequences of writing assessment," about the effects of writing assessment on diverse students.

Edel Reilly is Associate Professor in the Mathematics Department at Indiana University of Pennsylvania, where she teaches courses for elementary and middle level education majors. Her research focuses on mathematics and writing, middle level mathematics education, curriculum integration, and gender issues in mathematics and science. Dr. Reilly is the author of *A smart way to teach math*, which focuses on how to use interactive technology in the mathematics classroom. Her work on curricular integration has been published by the National Council of Teachers of Mathematics and the National Visual Literacy Association.

Marin S. Robinson is Professor and Chair of Chemistry and Biochemistry at Northern Arizona University. She teaches courses in organic chemistry, atmospheric chemistry, and scientific writing. Her research interests include atmospheric chemistry and chemistry-specific writing. She has co-authored publications on disciplinary writing, including *Write like a chemist: A resource and guide* (Oxford University Press, 2008) and a chapter in the edited volume *English language teaching textbooks: Content, consumption, and production* (Palgrave Macmillan 2014).

Fredricka L. Stoller is Professor of English at Northern Arizona University where she teaches in the MA TESL (Teaching English as a Second Language) and Ph.D. in Applied Linguistics programs. She is a co-author of *Write like a chemist* (Oxford University Press, 2008). Other publications on disciplinary writing have appeared in *English for Specific Purposes*, *Journal of English for Academic Purposes*, *Journal of Chemical Education*, and *English Teaching Forum*. Her professional interests include disciplinary writing, second language reading, project-based learning, and teacher training.

Book series preface

Language studies, science and engineering

It is with great pleasure that I introduce this first book edited by Mary Jane Curry and myself in the new series – *Language Studies, Science and Engineering* – from John Benjamins. The book series (and this book) developed over several years as part of an increased interaction between applied linguistics and the various fields of STEM. This book grew out of both Mary Jane's and my own work in collaborating with STEM professionals and applied linguists in disciplinary endeavors and as disseminators of this research experience through collaborative conference panels directed at generating interest in and understanding of the value of applied linguistics in science, technology, engineering, and math. The book brings together a range of researchers and research projects that exemplify the power and potential of using linguistic methodologies for questions in STEM. As such, this book is a perfect way to launch this book series.

The aim of the book series is to address the interface among language studies, science, engineering, and education. The series aims to bring together researchers from the fields of language studies and science/engineering education with the aim of generating new interdisciplinary knowledge. It is premised on the concept that science is of central importance in the twenty-first century and that research informed by linguistic knowledge can contribute to the description, understanding, education, and practice of science, technology, mathematics, and engineering. The goal of this series is to enhance educational and professional practices in the sciences and engineering through interdisciplinary interaction among language researchers, science and engineering educators, and scientists. The scope of this book series covers the range of potential contributions that language studies can make to the advancement of science, engineering, and educational practices in these fields including (but not limited to) discourse analysis, computational linguistics, conversational analysis, multimodal analysis, rhetorical analysis, and genre analysis.

The book series exists in order to disseminate work at the intersection of language studies and STEM. This first book summarizes some of the approaches by researchers working in this arena. If you are reading this book as a STEM professional, this book (and series) may offer you new directions for answering questions; if you are an applied linguist (or other language professional) this book may offer you new avenues for extending your methodological knowledge and the disciplinary contexts. My hope is that this book will excite, interest, and entice additional applied linguists and STEM professionals, interdisciplinary teams, and research projects to emerge and ultimately publish their results in this book series.

David Ian Hanauer

Professor, Graduate Program in Composition and TESOL, Department of English, Indiana University of Pennsylvania

Assessment Coordinator, PHIRE Program, Hatfull Laboratory, Pittsburgh Bacteriophage Institute, University of Pittsburgh, Pennsylvania

CHAPTER 1

Integrating applied linguistics and literacies with STEM education

Studies, aims, theories, methods, and forms

David I. Hanauer and Mary Jane Curry
Indiana University of Pennsylvania / University of Rochester

This chapter describes the impetus for this collection of papers and provides an overview of the ways in which applied linguistics and literacies can be integrated with the fields of science, technology, engineering, and mathematics (STEM). We first explore the value of applied linguistics and literacies approaches to researching STEM education. Through a review of the studies included in this book, we present their underpinning theoretical orientations, research methodologies, communicative forms, and aims. We propose a facilitative role for applied linguistics in terms of the broad educational agenda of STEM.

Introduction

Language and literacy are central to the conduct and communication of science, technology, engineering, and mathematics. Yet the workings of language and literacy in the activities of STEM fields are often obscured by a focus on scientific procedures and outcomes. Applied linguistics is an interdisciplinary field that offers tools with which to explore language usage in a range of professional settings (Pennycook 2001). In this book, we focus on applied linguistics as a nexus at which linguistic knowledge, understanding of a wide range of representational forms including written academic texts, commitment to empirical research methodologies, and real world problems meet. This book (and the projects presented within it) is a case in point. The aim of the book is to exemplify ways in which applied linguistics knowledge and expertise can be useful in researching, understanding, and improving issues that have arisen within the fields of science, mathematics, and engineering.[1] In the past few decades awareness has grown both within and outside STEM fields of the need for STEM practitioners and educators to understand the central role played by language and literacy in the activities of the disciplines – both those considered to be central to disciplinary activity as well as those typically seen as related to communication about disciplinary work to wider audiences (e.g. Bazerman 1988; Hyland 2004; Knorr-Cetina 1981; Latour & Woolgar 1986). Indeed, calls have come from within the STEM fields for helping students to understand the central role of language and literacy in science and to improve their ability to work with disciplinary language and genres (Hanauer, Frederick, Fotinakes, & Strobel 2012; Thomson 2003; Yore, Bisanz, & Hand 2003). Addressing this need, this book presents a collection of studies that exemplify the value of research on the ways that language and literacy function in the STEM fields and how these understandings are communicated to others.

This book is unusual in bringing together papers by applied linguistics and literacy researchers and STEM professionals and educators aiming to address disciplinary problems and explore the ways in which applied linguistics can facilitate solutions to these issues. The aim of this chapter is to offer the "big picture" view of what the use of applied linguistics in STEM education can mean and to explore the ways in which this form of interdisciplinary collaboration and research can be conducted. This discussion, as with the whole book, is directed at promoting increased integration of applied linguistics and literacies with STEM education through the exemplification of existing projects. Our assumption is that by looking at work that has been done and carefully considering the research methods used, collaborations undertaken, and emerging understandings, a much better sense of how applied linguistics and STEM can interact will be facilitated. Ultimately, our aim is to promote increased understanding for both STEM professionals and applied linguists of the mutual benefits of drawing these fields together and working together.

Integrating applied linguistics and literacies with STEM education

What do applied linguistics and literacies have to offer STEM professionals? Broadly, the answer to this question is that applied linguistics provides tools with which to answer many of the questions that arise from the educational work con-

1. Although we use the STEM acronym for convenience, this book does not include a focus on technology per se.

ducted in STEM fields. This is particularly true for those situations in which language, communication, and literacy in all their varied forms are an inherent part of professional practices and where aspects of these practices need to be made accessible to novices. An example from this book is Hanauer's work on laboratory notebooks. Keeping laboratory notebooks is a prevalent literacy practice within laboratory settings. Understanding what laboratory notebooks consist of, how they are used, and how they relate to scientific thinking gives insight into how students can be socialized into becoming scientists as well as offering options for authentic methods of assessing scientific inquiry. From the applied linguistics and literacies perspective, a laboratory notebook is both a literacy practice and a product: as such, a range of established approaches exist for investigating and understanding this genre, the results of which help us understand both learning and assessment in authentic (laboratory) contexts. Another example is Reilly's work on the use of writing in the mathematics classroom as a way of enhancing female students' involvement with mathematics. Once again, the issue of increasing girls' participation in and understanding of mathematics comes from the realm of education in a STEM discipline, while the problem is addressed through research conducted on writing from the language sciences.

A more specific answer to the question – What does applied linguistics have to offer STEM professionals? – is that applied linguistics offers three types of knowledge: knowledge of linguistic, rhetorical, and other representational forms; knowledge of quantitative and qualitative research methodologies; and knowledge of theoretical approaches to language, communication, and literacy. All of these types of knowledge can be brought to bear on issues that arise within the realm of STEM education and professional work.

Looking across the studies presented in this book, it is clear that a particular theoretical orientation underpins all the chapters. It is expressed in different ways and using varied terminological orientations such as English for Special Purposes (ESP), Writing in the Disciplines (WID), Writing across the Curriculum (WAC), genre theory, and discourse analysis. But what is shared by contributors is the understanding that language and literacy are specific to disciplines. In other words, the investigation, analysis, and discussion of different literacy products, multimodal objects, or oral interactions is contextualized within a framework of disciplinary action that underpins and explicates these communicative components. The assumption is that the communicative products of STEM fields result from the aims and activities of professionals and students within these fields. They are inexplicable without carefully understanding what those in the disciplines are trying to achieve and the ways in which they work. In this sense, any useful applied linguistic analysis is framed within a conceptualization of the discipline, including its conventions of communication and real-world functions.

Directly tied to this theoretical understanding of disciplinary-specific communication is a closely related sociocultural theory of learning that values researchers' understandings of the participation of members of a particular social practice (Lantolf 2000). Disciplines are by definition constructed by active and participating members who are usually organized on a continuum along expert/experienced to novice/student members. Sociocultural theories of learning place importance on an understanding of community and how people can become members of this community (e.g. Gee 2012). An applied linguistics approach to sociocultural theory explores the communicative means by which members of a social practice demonstrate their engagement within the community, at whatever point on the novice/expert continuum they may be. Speech and writing are often very explicit markers of community membership and degree of expertise. If a person can produce a linguistic product that is highly valued by the community, it is assumed that she is a member with standing in that community. This perspective underpins the work presented in this book, with chapters exploring ways of understanding disciplinary, community-based literacies and providing students/novices with ways of engaging with and understanding these literacies.

The research methodologies presented in this book are related to the sociocultural theoretical orientation that underpins the studies it presents. One aim of these chapters is to explore contextualized, disciplinary-specific communication. The foundation for such research is authentic disciplinary practices and products. The methodologies used to understand these contexts broadly combine two elements: (1) an explicit attempt to describe communicative products and social practices; and (2) an attempt to explore participating community members' understandings of these products and practices. In describing disciplinary products, practices, and understandings, these studies may use methodologies of document analysis, interview, observation, conversational analysis, discourse analysis, and survey. Many of the chapters included here involve more than one of these methods in an attempt to capture various aspects of the role of communicative products within disciplinary contexts. The majority of studies use systematic qualitative approaches and take an ethnographic orientation (although this is not always explicitly acknowledged), aiming for rich contextualized data that enables analysis of participant understandings of their practices and contexts.

Applied linguistics and literacies provides a broad set of tools with which to categorize and analyze communicative products. These include a range of linguistic categories, categories for exploring conversational interaction, move analysis for textual and oral texts, and coding systems for analyzing documents and written texts. In some cases, these categorization systems are applied through close human-based analysis, but there is also the option (as discussed in Stoller and Robinson) for automating these systems using computational linguistics. In this case, once a communicative product has been categorically analyzed, frequencies can be computed and descriptions provided. Applied linguistics is particularly good at developing systematic approaches of this type. As exemplified in this book, applied linguistics has adapted well to applying established and new categorization systems to STEM fields by offering tools that provide explicit, repeatable, and valid descriptions of communication in STEM.

This book exemplifies how applied linguistics studies communicate within the disciplinary context of STEM. Having reviewed the theoretical orientations, methodologies, and communicative forms presented in the studies in this book, it is important now to explicate the aims of the studies. In this way we can really get at the ways in which applied linguistics and literacies can be integrated with STEM agendas for education. Broadly, all the studies here fall into one of four aims: identifying disciplinary practices, improving disciplinary instruction, facilitating professional socialization, and enhancing professional communication. All of these aims have an educational focus.

Identifying disciplinary practices is a foundational, yet often overlooked, aim of applied linguistics and literacies studies. Participants inside a disciplinary practice are often those in the worst position to be able to identify and analyze the component parts of their activities, a step necessary for both understanding and teaching these practices. For example, both chapters in the Research Studies in Engineering section (Gimenez and Curry) illustrate the importance of identifying and explicating the values and activities that underpin the social practices of engineering communication.

Improving instruction – whether explicit or implicit teaching – is a basic and prevalent aim across the chapters in this book. The idea is to use the knowledge generated from applied linguistics research on teacher interaction, student writing, student talking, and professional communications to improve instruction in both formal and informal settings. In studies such as Lerner and Poe, Stoller and Robinson, Gimenez, and Moreira and Latas, the movement from analysis to instruction design is immediate and could directly inform the design of materials and courses. Drawing on studies such as Choppin, Reilly, Hanauer, and Curry, instruction could be improved through increased understanding of the interactions and products used in education.

In summarizing the ways in which studies in this book integrate applied linguistics and literacies with STEM education, the following foundational propositions can be articulated:

1. Communication in STEM fields is a contextualized, disciplinary-specific activity involving particular conventions and real world functions.
2. Communication in STEM disciplines is multimodal, including spoken, written, pictorial, and physical representations.

3. Analyzing communication in STEM involves exploring communicative forms, processes of production, and disciplinary practices.
4. Socialization in STEM disciplines involves learning expertise through engaging in production of the often-multimodal, communicative forms of that discipline.
5. Membership in a STEM discipline is manifest through communicative participation.
6. Applied linguistics and literacies research methodologies provide tools for investigating, categorizing, and theorizing the various aspects of communication in STEM.
7. Educational approaches in STEM are enhanced when they are based on research-based understandings of the forms of communication used in both professional STEM settings and educational programs.

Book structure and chapter summaries

Following this introductory chapter, the book is divided into three sections that present research studies in science, engineering, and mathematics. These sections are followed by a final chapter discussing an established funding program aimed at improving education in STEM disciplines. The Research Studies in Science section opens with a chapter from Stoller and Robinson that reports on the Write Like a Chemist project, in which faculty members in chemistry and applied linguistics collaborated to analyze the language of chemistry in four types of disciplinary writing. Findings of this collaboration were used to design instructional materials and to develop a discipline-specific course aimed at improving students' disciplinary literacy. Usefully for the opening of this section, the chapter illustrates the contributions of various subfields of applied linguistics, including English for Specific Purposes, discourse analysis, corpus linguistics, and language teaching pedagogy. Next comes Hanauer's chapter, already mentioned, in which he describes an exploration of the genre characteristics of laboratory notebooks that enabled him to organize these characteristics according to social functions, models of notebook completion, and types of entries students made in the notebooks. He finds the laboratory notebook a multifaceted literacy object that functions as a scientific document for recording laboratory activities, a learning tool for acquiring laboratory protocols, and a personal journal for enhancing students' scientific engagement. The final chapter of the science section by Lerner and Poe presents a follow-up study from writing scholars Poe, Lerner, and Craig's (2010) book, *Learning to Communicate in Science and Engineering*. They present three case studies of students whose academic trajectories shifted over the length of their undergraduate study, focusing on the relationship between students' developing scientific knowledge, changing identities, and the role of writing in these experiences and changes.

The Research Studies in Engineering section opens with Gimenez's chapter contributing the findings of a qualitative study of four engineering departments at one British university, documenting sub-disciplinary variation in the communicative practices, values, and ideologies across these departments. This study reaffirms the importance of seeing disciplinary communication practices as locally and contextually situated. Curry's chapter, reporting on findings from a longitudinal ethnographic study, focuses on the notion of invention in academic writing for publication, that is, the moments when research engineers identify the research results they want to present and decide on the arguments they want to make in an article. It argues that in engineering writing for publication, graphics play a crucial heuristic role at the moment of invention.

All three chapters in the Research Studies in Mathematics section take a broad look at learning in the K-12 classroom, though each chapter has a different focus. Moreira and Latas write about an ethnographic study they conducted in a coastal town in Portugal with many immigrant and low-income children where surfing became the cultural context for teaching mathematics. They argue that connecting such cultural practices with mathematical tasks enhances students' ability to understand and articulate mathematics. Shifting the focus from helping students produce disciplinary talk to listening to students talk about mathematics, Choppin uses classroom discourse data to analyze different types of listening by teachers. Grounded in the conviction that listening is a key to understanding what students know and don't know about mathematics, Choppin illustrates three types of listening – evaluative, interpretive, and hermeneutic – and demonstrates the particular pedagogical value of the latter. Highlighting the need to bring more girls into mathematics, Reilly's chapter presents a mixed-method study in which she analyzes the reflective writing by middle school students and surveys these students on their perspectives on the value of writing. She concludes that writing can be used as a way of bringing girls into mathematics, providing sample statements from students that illustrate this claim.

The book concludes with a description of a unique program of the Howard Hughes Medical Institute (HHMI) to support the educational work of established scientists. Co-authored by HHMI Program Officer Bauerle, an applied linguist, Hanauer, and a scientist, Hatfull, it provides the example of the collaboration between Hanauer and Hatfull in The Phage Hunting Integrating Research and Education (PHIRE) program. This collaboration led to the creation of interdisciplinary education projects that furthered the HHMI program's pedagogical mission. The chapter is included in order to highlight the recognized value of such collaborations and to encourage other cross-disciplinary collaborations to be undertaken, documented, and ideally, to be funded.

The editors are grateful to Indiana University of Pennsylvania graduate student Fang-Yu Liao for help in compiling the index.

References

Bazerman, C. (1988). *Shaping written knowledge: The genre and activity of the experimental article in science.* Madison, WI: University of Wisconsin Press.

Gee, J.P. (2012). *Social linguistics and literacies: Ideology in discourses* (4th ed.). London: Routledge.

Hanauer, D., Frederick, J., Fotinakes, B., & Strobel, S. (2012). Linguistic analysis of project ownership for undergraduate research experiences. *CBE-Life Sciences Education, 11*, 378–385.

Hyland, K. (2004). *Disciplinary differences: Social interactions in academic writing.* Ann Arbor, MI: University of Michigan Press.

Knorr-Cetina, K. (1981). *The manufacture of knowledge: An essay on the constructivist and contextual nature of science.* Oxford: Pergamon Press.

Lantolf, J. (2000). *Sociocultural theory and second language learning.* Oxford: Oxford University Press.

Latour, B., & Woolgar, S. (1986). *Laboratory life: The construction of scientific facts.* Princeton, NJ: Princeton University Press.

Pennycook, A. (2001). *Critical applied linguistics: A critical introduction.* Mahwah, NJ: Lawrence Erlbaum Associates.

Poe, M., Lerner, N., & Craig, J. (2010). *Learning to communicate in science and engineering.* Cambridge, MA: The MIT Press.

Thomson, N. (2003). Science education researchers as orthographers: Documenting Keiyo (Kenya) knowledge, learning and narratives about snakes. *International Journal of Science Education, 25*(1), 89–115.

Yore, L., Bisanz, G.L., & Hand, M. (2003). Examining the literacy component of science literacy: 25 years of language arts and science research. *International Journal of Science Education, 25*(6), 689–725.

PART I

Research studies in science

CHAPTER 2

Drawing upon applied linguistics to attain goals in an interdisciplinary chemistry–applied linguistics project

Fredricka L. Stoller and Marin S. Robinson
Northern Arizona University

Students in the fields of science, technology, engineering, and mathematics (STEM), like students in other tertiary-level areas, benefit from learning to read and write in their disciplines as they progress in their studies. This chapter reports on a successful collaborative project that brought together faculty in chemistry and applied linguistics to analyze the language of chemistry in four types of disciplinary writing (i.e. genres). Findings were used to design instructional materials and develop a discipline-specific course, to be taught by chemistry faculty, aimed at improving students' disciplinary literacy skills. The focus of the chapter is on the contributions of various subfields of applied linguistics to project goals, emphasizing the contributions of English for Specific Purposes, discourse analysis, corpus linguistics, and language teaching pedagogy. Interdisciplinary partnerships, such as the one showcased in this chapter, are particularly fruitful when disciplinary-language issues are at stake. STEM faculty and faculty in language-related disciplines can combine their different but complementary resources to achieve goals that might not be possible to attain independently. With a heightened understanding of disciplinary language and the literacy practices specific to particular STEM fields, meaningful instructional materials and discipline-specific classes can be developed to give students access to and control of the literacy practices and genres valued by STEM fields.

Introduction

It is not often that science, technology, engineering, and mathematics (STEM) faculty find themselves collaborating with applied linguistics faculty on more than institutional committees and faculty senate. The scarcity of such collaboration exists despite the fact that many tertiary-level institutions and funding agencies look favorably upon interdisciplinary endeavors. Nonetheless, partnerships between STEM and applied linguistics faculty, the latter often housed in English departments, can be particularly fruitful when disciplinary-language issues are at stake. And language issues do emerge as STEM students progress in their university studies (and beyond), in part because literacy development is typically neglected throughout college years (see National Commission on Writing, 2003). For example, students oftentimes experience difficulties transitioning from general academic English to the written and oral conventions of their STEM disciplines. Similarly, students' experience reading comprehension problems as they move from introductory textbooks in lower-division classes to the genres valued in their disciplines in upper-division classes and graduate school (e.g. peer-reviewed journal articles, technical and design reports, and theses). These reading and writing difficulties stem from the fact that each type of writing (i.e. genre) is characterized by its own content, form, style, audience, and purpose. Furthermore, when students are assigned to write in these genres, they often are unaware of the organizational conventions, formatting expectations, vocabulary choices, and content (what to include and exclude) that would make their writing acceptable in the eyes of the discipline. For many STEM students, the specialized vocabulary that they need to master (along with associated abbreviations and acronyms) becomes a formidable task as they progress in their studies. These are but just a few of the language-related issues that STEM students face in their university years.

When STEM and applied linguistics faculty join forces to solve content-area language issues such as these, solutions that might be challenging to attain independently become real possibilities. In such partnerships, STEM faculty contribute their knowledge of the content area, their familiarity with content-area curricula and the literacy demands of different classes, and their personal experiences with the conventions of discipline-specific written and oral genres (e.g. design reports, journal articles, posters, research-group presentations, research proposals). To complement STEM faculty expertise, applied linguists contribute at least the following to the partnership: language expertise; knowledge of approaches to literacy instruction; familiarity with second language student populations and the language-related challenges that they face in university classrooms; and a commitment to solving language-related problems, the latter a distinguishing attribute of the field. When discipline-specific literacy, in the broadest terms, is of concern, interdisciplinary collaboration is a logical path to pursue.

This chapter highlights a successful interdisciplinary endeavor that brought chemists and applied linguists together to design a discipline-specific course aimed at improving chemistry students' disciplinary reading and writing. The outcomes of the multi-year Write Like a Chemist project (described below) were achieved as a result of the merging of two diverse areas of specialization, chemistry and applied linguistics. We begin the chapter by providing background on the project and then focus on the contributions of various subfields of applied linguistics to project goals, placing emphasis on the contributions of English for Specific Purposes, discourse analysis and corpus linguistics, and language teaching pedagogy. Although the contributions of each subfield are discussed separately, in actuality, they overlap and inform one another in important ways. The chapter concludes with implications for other STEM areas.

Background

The Write Like a Chemist project was conceived as a response to a Northern Arizona University (NAU) mandate to address the writing needs of third-year undergraduate students. Departments were given the option of developing third-year writing-intensive courses of their own or requiring students to take English Department courses. The Chemistry Department chose to develop its own course, but not by itself. The chemistry faculty member who spearheaded the course-development process initiated a "cross-disciplinary alliance" (Wardle 2004) with an applied linguist in the English Department. The course-development team expanded, at different times, to include two graduate students in applied linguistics and a post-doctoral associate in chemistry. Over time, the project took on more ambitious goals and course materials were piloted at 16 U.S. universities over a two-year piloting cycle. The pilot involved 30 additional chemistry faculty – as piloters, reviewers, and external evaluators – as well as an applied linguist from the English Department, who served as the director of project assessment.

The Write Like a Chemist project, driven from the onset by pedagogical aims, was directed toward two primary audiences: chemistry students and chemistry faculty. Chemistry students (native and nonnative English speakers) were targeted at the point in their university studies when they are transitioning to disciplinary reading and writing. Chemistry faculty members, who are not typically trained to teach disciplinary writing, were targeted as well.

As part of the project, four chemistry genres were selected for textual analysis and explicit instruction: the journal article, research proposal, and the interconnected conference abstract and scientific poster. The results of our textual analyses, rather than the use of intuitions about the language of chemistry and generalizations about the language of science, were transformed into instructional materials and a read-analyze and write pedagogical approach to literacy-skills instruction. Our efforts led to two tangible outcomes: a textbook (Robinson, Stoller, Costanza-Robinson & Jones 2008) and its companion website ⟨http://www.oup.com/us/writelikeachemist⟩.

Contributions of applied linguistics subfields to the Write Like a Chemist project

Aspects of the Write Like a Chemist project were informed by various subfields of applied linguistics. These subfields, with their own methodologies and paradigms, made notable contributions to the project. Together they led to a heightened understanding of the language of chemistry in the four target genres. Furthermore, they influenced the design of a pedagogical approach to discipline-specific literacy instruction and the development of corresponding instructional materials, written for our two target audiences. In the sections that follow, we introduce the subfields of applied linguistics that had the greatest influence on the project. At different points in our discussion, we showcase a sampling of "artifacts," with linkages to those subfields, that have been well received by chemistry faculty and students.

English for Specific Purposes

English for Specific Purposes (ESP), a subfield of applied linguistics, played a foundational role in the Write Like a Chemist project. ESP is grounded in the study of language, discourses, and contexts-of-use and then the application of findings to pedagogical practice (Belcher, Johns, & Paltridge 2011). ESP practitioners typically strive to provide language instruction that is practical and responsive to the goals and realities of particular groups of students, thus ESP approaches are often referred to as learner centered (Basturkmen 2006, 2010; Belcher 2009; Dudley-Evans & St. John 1998). To determine students' needs, the field places importance on needs analysis and, as part of the needs analysis, on interactions with experts in the target discourse community (Basturkmen 2010; Belcher 2009; Tribble 2010). ESP researchers and practitioners make use of needs-assessment data and the results of linguistic analyses in ESP course- and materials-design, task-development, teacher-training, and assessment efforts (Douglas 2000).

ESP recognizes that linguistic variation exists not only across disciplines, professions, and occupations but also across genres within them (e.g. Bhatia 2004; Gunnarsson 2009; Hyland 2004b; Hyland & Bondi 2006; Johns 1997, 2007; Swales 1998; Swales & Feak 2011; see also the journals *English for Specific Purposes; Across the Disciplines: Interdisciplinary Perspectives on Language, Learning, and Academic Writing;* and *Journal of English for Academic Purposes*). In fact, ESP attributes the success of students engaged in English for specific purposes in part on their access to and control of disciplinary genres (Basturkmen 2010; Bhatia 2004; Nesi & Gardner 2012; Tardy 2009). To identify the defining characteristics of the language and valued genres of particular disciplines, ESP researchers make use of advances in and methodologies associated with two other applied linguistics subfields, specifically discourse analysis and corpus linguistics. Some of the ways in which the Write Like a Chemist team made use of these subfields are described in the section that follows.

Discourse analysis and corpus linguistics

From the beginning of the Write Like a Chemist project, the interdisciplinary team realized that it needed to examine the language of chemistry in the four genres targeted for instruction. We needed to understand how chemists use and organize language to accomplish their goals in each genre. Therefore, we turned to the subfields of discourse analysis and corpus linguistics to guide our investigations of the language of four chemistry genres. According to Paltridge (2006), *discourse analysis* involves the study of language – beyond single words, clauses, phrases, and sentences – for the purpose of identifying the linguistic patterns that exist across texts (and sections within them). Discourse analysts are interested in how people organize what they write (and say) and what relationships exist between language and the social and cultural contexts in which it is used. Swales' "move analysis" (1990, 2004), a component of discourse analysis, allows for an in-depth investigation into textual organization and the functions of each "move" that writers make as they progress through the larger text. *Corpus linguistics* involves the empirical investigation of language through computer-based analyses of large collections of texts, known as corpora (Biber, Connor, & Upton 2007; Biber, Conrad, & Reppen 1998; Bowker & Pearson 2002; Hunston 2002). Corpus research findings are commonly used to inform ESP instructional practices (e.g. O'Keeffe, McCarthy, & Carter 2007; Reppen 2010). It is worth pointing out that chemists have responded favorably to the objectivity of findings about the language of chemistry that have stemmed from discourse and corpus analyses.

By making use of discourse-analysis and corpus-linguistics tools, the Write Like a Chemist team examined authentic samples of chemistry discourse in four genres. We analyzed *full* journal articles, research proposals, conference abstracts, and posters and the *sections* within them (e.g. the Abstract, Introduction, Methods, Results, Discussion, and Conclusion in journal articles). One of the goals of these analyses was to ascertain how varied elements of writing coalesce in order for a writer's objectives to be achieved. These writing elements include the following:

- lexico-grammatical features (e.g. tense-voice combinations, word usage, grammatical patterns, common lexical bundles)
- organization (across the full document and within sections)

- communicative function (e.g. to provide relevant background information in a journal article Introduction, to refer to a figure or table in a journal article Results section, to share corroborating or conflicting results in a journal article Discussion section)
- disciplinary conventions (e.g. how to report numbers and units, format tables)
- content and the ways in which it is presented in text and graphics
- mechanics (e.g. use of punctuation)

We also explored how new topics are introduced, transitions are accomplished, ideas are developed, and concise writing is achieved. In addition, we investigated how the choice of different language features (e.g. pronouns, tense-voice combinations, vocabulary) affect the discourse, how and where authors hedge (to show restraint in their claims; see Hyland 1998), and how authors use language to reach different audiences, even within the same document (e.g. *scientific* audiences in journal article Introductions; *expert* audiences in journal article Methods sections).

We engaged in other analyses to dispel commonly held beliefs among chemists. For example, in light of the common belief that scientific discourse does not use first-person pronouns (e.g. I, we), we analyzed historical changes in the use of the first-person plural pronoun "we" in three chemistry journals over three time periods. From that analysis, we learned that, in fact, chemists do use "we" (Figure 2.1) and do use it predictably at certain points in the genre.

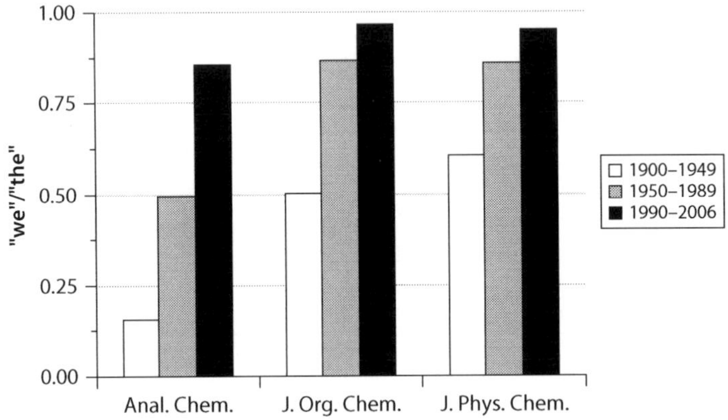

Figure 2.1. Fraction of documents using *we* once over three time periods in *Analytical Chemistry, Journal of Organic Chemistry*, and *Journal of Physical Chemistry*, determined using the American Chemical Society Journal Search. From Robinson et al. (2008, p. 149)

Similarly, we were able to dispel another widespread myth, this one about passive constructions: Many chemists (and writing teachers) are under the impression that the passive is "a weak language construction" and used rarely among chemists. Based on our objective analyses of peer-reviewed journal articles, we were able to point out where (Figure 2.2) and for what purposes the passive is used by chemists. We also explored chemistry faculty pet peeves – such as when students omit a space between a number and its unit (e.g. 10 mL, not 10mL), incorrectly use lists rather than full sentences to describe materials, and select the wrong word in an easily confused word pair (e.g. affect/effect, its/it's, comprise/compose) – to assist us in designing instructional materials that chemistry students would find particularly helpful.

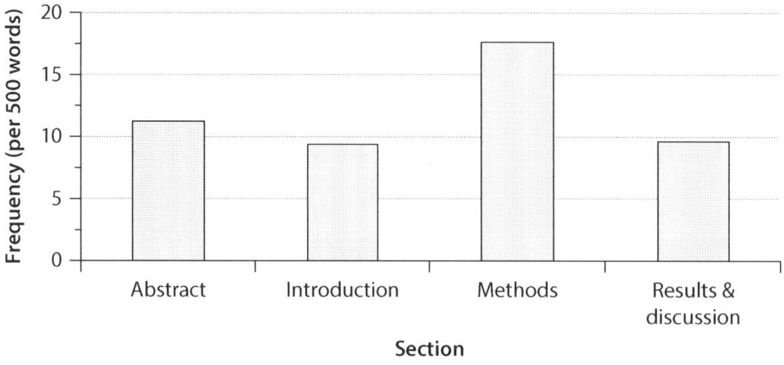

Figure 2.2. Frequencies of the use of passive voice in sections of chemistry journal articles (expressed as the number per 500 words), determined through a computer-based analysis of 60 articles. From Robinson et al. (2008, p. 99)

Early in the project, a project-specific corpus of chemistry journal articles was developed for analyses. It comprised 200 full-length refereed chemistry journal articles (for a total of 991,606 words) and 240 sections of refereed chemistry journal articles, including 60 Abstracts, Introductions, Methods, and Results and Discussions sections (for a total of 297,407 words). The corpus (with a total of 1,289,013 words) was used to examine numerous features of writing, including those itemized in Table 2.1. Later, we learned to use the extensive American Chemical Society (ACS) Journal Search data base ⟨http://pubs.acs.org⟩ for similar purposes (Robinson, Stoller, & Jones 2008). At the time of our work, the ACS's 1996-to-Current Issue database included more than 395,000 articles, from 45 ACS journals.

Following the seminal work of Swales (1990, 2004) and adherents of the ESP school of genre analysis (Bawarshi & Reiff 2010), we compiled additional sets of authentic examples of our target genres for hand-tagged analyses of the "moves" (i.e. organizational features and their functions) in each genre. Precedents in applied linguistics for discipline-specific and cross-disciplinary move analyses

Table 2.1. Sample features of chemistry writing investigated with discourse-analytic and corpus linguistics tools

Sample features of writing	Sample findings
Collocations (i.e. words that go with other words)	*ab initio*, when used as a two-word modifier, is often followed by these words: molecular dynamics, simulations, calculations, study
Common verbs following *we* in the "fill-the-gap statement" toward the end of a journal article Introduction (In this work, we …)	*Present tense* is commonly used with these verbs: carry out, demonstrate, describe, develop, employ, present, propose, report, show, use; *Past tense* is commonly used with these verbs: analyzed, calculated, chose, determined, employed, examined, focused on, found, investigated, measured, solved, studied, synthesized
Communicative function of verb tense-voice combinations	*Present-active* is used, e.g. to introduce the research area in a journal article Introduction; *Past* (*active* or *passive*) is used, e.g. to share corroborating or conflicting results of others' work in the Discussion section of a journal article
Conciseness (highly valued in chemistry)	*because* (vs. due to the fact that, based on the fact that); *although* (vs. despite the fact that, in spite of the fact that); *is/are* (vs. is/are known to be); abbreviations and parentheses (e.g. … methanol (1 mL) vs. … one milliliter of methanol)
Hedging (i.e. the use of words to indicate some degree of caution, tentativeness, or restraint)	data *suggest* (vs. *prove*); results offer *evidence* (vs. *proof*); hedging is used often in journal article Discussion sections and in the presentation of timelines in research proposals
Lexical frequencies	*As shown in Figure* (the most common four-word bundle in our corpus); *In the present study, we* (a common phrase used to transition to the final move in a journal-article Introduction)
Nominalizations (as means to achieve conciseness)	Common nominalizations in chemistry writing include *absorption, conductivity, diffusion, excitation, luminescence, oxidation, reactivity*
Two-word modifiers	noun + adjective (*time-dependent*), adjective + noun (*long-term*), adjective + participle (*fast-growing*), noun + particle (*phospholipid-coated*), noun + noun (*solvent-solute*)
Word choice	*monitor, measure, investigate* versus *look into*
Word usage	*affect* versus *effect, adsorb* versus *absorb, since* versus *because*

are many (Biber, Connor, Upton, & Kanoksilapatham 2007), including those that focus on sections of genres in STEM fields (e.g. Basturkmen 2012; Bruce 2009; Peacock 2002; Samraj 2002; Swales 1990, 2004; Williams 1999) and the few that focus on full-length exemplars in STEM (Kanoksilapatham 2005, 2007; Lin & Evans 2012; Nwogu 1997; Posteguillo 1999). For the Write Like a Chemist project, move-analysis findings were converted into "move structures akin to flow charts" to showcase the conventional organization (and corresponding functions) of our target genres and their typical sections (Stoller & Robinson 2013). To illustrate, we include a move structure for the Methods section of a research proposal in Figure 2.3. Move structures such as these have proven effective as instructional tools for "newcomers" to discipline-specific reading and writing.

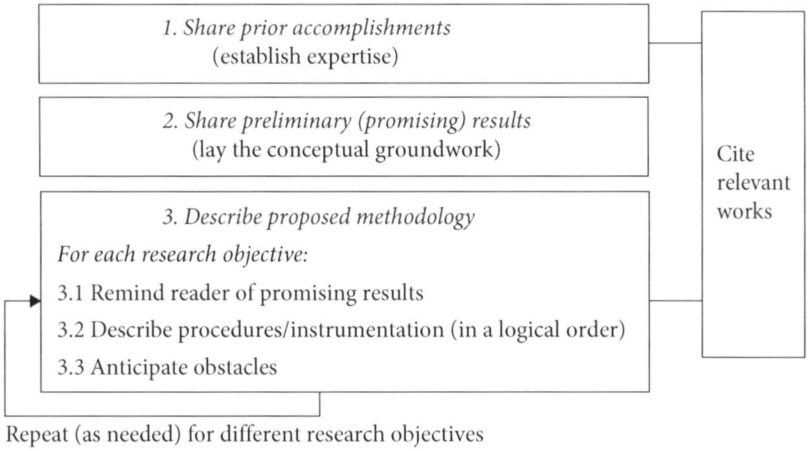

Figure 2.3. Organization of the methods section of the project description in a research proposal (from Robinson et al. 2008, p. 437)

Language teaching pedagogy

The Write Like a Chemist team drew upon various applied linguistics areas concerned with language teaching pedagogy. Specifically, the applied linguistics literature on course and syllabus design influenced many of our course-development decisions. Similarly, the vast literature (and research) on language teaching practices, more generally, and writing instruction, more specifically, contributed to the project in various ways. Although applied linguists typically focus on second, foreign, or bilingual student populations, the Write Like a Chemist team adapted features of that language teaching pedagogy for the purposes of developing a university-level, discipline-specific literacy course for native and nonnative speakers of English.

Course and syllabus design. With regard to course and syllabus design, the Write Like a Chemist team made use of numerous practices endorsed by the applied linguistics community, including conducting needs and situational analyses; articulating goals, objectives, and learner outcomes; specifying course content and its sequencing; and planning for pre and post evaluation as well as periodic student, peer, and teacher assessments (Brown 1995; Ferris & Hedgcock 2013; Nation & Macalister 2010; Richards 2001). As part of our needs analysis, we designed, piloted, refined, and administered pre- and post-assessment measures (Stoller, Horn, Grabe, & Robinson 2005, 2006) and considered what students would have to be able to do in the future (with regard to our four target genres). Integral to our situational analysis was our consideration of higher-education institutional factors, American Chemical Society accreditation issues, the typical undergraduate chemistry curriculum (which often has little wiggle room for writing instruction), chemistry faculty preparation to teach writing, learner factors, and the fact that few discipline-specific instructional materials were available to help us reach our project aims. Course goals, objectives, and learner outcomes were articulated based on an understanding of learners' present and long-term literacy needs (Richards 2001); they were reviewed by chemistry faculty who served as external evaluators for the larger project.

Tribble (2010) has emphasized that writing teachers require not only "a clear understanding of … students' needs" but also the ability "to develop a program that will enhance [students'] capacity to write the texts which matter to them" (p. 162). A process approach to genre-based instruction (Paltridge et al. 2009; see also Cheng 2006; Hyland 2004a; Tardy 2009; Wingate 2012) was drawn upon to form the foundation of the Write Like a Chemist course. With genre as the guiding principle for course and syllabus design, we were able to address students' discipline-specific reading and writing needs (e.g. Hirvela 2004; Johns 1997; Plakans 2009). To ease learners into the language of the discipline, we provided opportunities for students to use and build on prior knowledge (Tribble 2010), allowing them to transition from *known* genres (e.g. everyday genres, such as used car advertisements, and familiar chemistry-specific genres, such as the Material Safety Data Sheet) to *new* genres (i.e. our target genres). Other aspects of the Write Like a Chemist instructional approach, which are drawn from various applied linguistic traditions, are described in the section that follows.

Language teaching practices. In line with our commitment to give students access to valued chemistry genres, a read-analyze and write approach to genre-based instruction was developed (Robinson & Stoller 2007), whereby students read (and reread) authentic texts from the target genre, engage in scaffolded genre-analysis activities, and then write (and rewrite) their own work following predominant disciplinary conventions. Instructional scaffolding, which involves teacher guidance, teacher monitoring, and a gradual release of responsibility to students (Paltridge et al. 2009), is an important component of the approach. The approach also acknowledges and builds upon the connections between reading and writing that are critical for literacy-skills development (Grabe & Zhang 2013). With explicit instruction, repeated exposures, discussion, practice, feedback, and time, students gradually develop an understanding of valued disciplinary genres and their layers of complexity (Tardy 2009).

Even though we cannot predict exactly what students will be reading and writing in their futures (Johns 1997, 2007), Write Like a Chemist instruction was designed to provide students with analytic tools that would stand the test of time and allow students to determine disciplinary expectations independently in the future. As one example of such analytic tools, tasks were designed that required chemistry students to use the ACS Journal Search, just as course developers did, to discover textual attributes of target genres on their own (Robinson, Stoller, & Jones 2008). Other genre-analysis tasks (such as those incorporated into Robinson et al. 2008, and those endorsed by Hyland 2004a; Johns 1997; Paltridge 2001; and Swales & Feak 2012; among others) are similarly transferable to other settings.

Central to the process approach to genre-based instruction is scaffolding, noted above. By means of teacher modeling and guided practice, followed by learners' independent construction of new texts, students gradually become acquainted and comfortable with discipline-specific practices. Additional pedagogical practices commonly endorsed by "blue collar" applied linguists (i.e. those involved in and concerned with real-world classroom contexts, see Ferris 2005) were integrated into the Write Like a Chemist approach. These include:

- Student pre-writing, drafting, revising, editing, peer review (simulated and authentic), and reflection, all aspects of the writing-process approach (Ferris & Hedgcock 2013)
- Consciousness raising among students about audience and purpose and the ways in which they influence vocabulary choice, determine content (inclusion or exclusion), and shape organization, among other writing features (Hyland 2004a; Paltridge 2001; Swales & Feak 2012)
- The use of model texts (both in their entirety and excerpts) to showcase disciplinary writing conventions (Paltridge 2001; Paltridge et al. 2009; Swales 1990; Tardy 2009). Through the use of multiple model texts, rather than single examples (Tribble 2010), instructors can introduce students to the structure and main features of the target genre, in addition to common variations, before moving students toward production of their own written work.
- The development of holistic and analytic rubrics to guide chemistry faculty in providing meaningful feedback to students about their writing (Ferris 2011; Ferris & Hedgcock 2013; Grabe & Kaplan 1996; Paltridge et al. 2009; see also Stoller, Horn, Grabe & Robinson 2005, for a description of the process of creating and validating the rubrics for the Write Like a Chemist project).
- A focus on active student engagement, that is, learning by doing through meaningful tasks (see Willis & Willis 2007), rather than directing students to simply read about disciplinary writing

Alongside the needs of students are the needs of instructors (Tribble 2010). Because chemistry faculty are not typically trained to teach writing, it was important that our project also develop instructors' guidelines and teaching tips (rather than just an answer key). As part of our guidelines, we included not only general suggestions for teaching writing but also an overview of select features of our approach that would likely be new for chemistry faculty (e.g. move structures, guided peer review, reflective writing). In our teaching tips, we offered suggestions for, as examples, how to (a) deal with students' apprehensions about writing; (b) build students' confidence; (c) help students overcome writer's block; (d) build on chemistry students' attraction to objectivity (unlike the subjectivity experienced in some liberal studies classes, including more traditional writing classes); and (e) manage mixed-ability classes.

Implications for other STEM areas

In this chapter, we have showcased an interdisciplinary project (chemistry–applied linguistics) that drew upon several applied linguistics subfields, including English for Specific Purposes, discourse analysis, corpus linguistics, and language teaching pedagogy. The contributions of these subfields, paired with the subject-area expertise brought in by chemistry team members, allowed for detailed analyses of four chemistry genres and the development of a university-level discipline-specific literacy course, and corresponding materials, that could be taught by chemistry faculty to chemistry students. Project successes (in terms of both process and outcomes) are partially attributable to the partnership that was established between chemists and applied linguists.

As an extension, we suggest that STEM faculty concerned about the various linguistic challenges faced by their students consider partnerships with applied linguistics faculty. By means of such collaborative efforts, STEM and applied linguistics faculty can combine their resources (i.e. their areas of expertise, along with the methodologies and paradigms associated with each of their fields) to make gains that might not be possible if they worked separately. The real beneficiaries of such collaboration are students who need access to disciplinary practices and genres to become members of those discourse communities (Bhatia 2004).

Acknowledgments

We thank the National Science Foundation (NSF) for supporting the Write Like a Chemist project with grants DUE 0087570 and 0230913. Opinions, findings, conclusions, and recommendations expressed here are those of the authors. They do not necessarily reflect the views of the NSF.

References

Basturkmen, H. (2006). *Ideas and options in English for specific purposes*. Mahwah, NJ: Lawrence Erlbaum Associates.

Basturkmen, H. (2010). *Developing courses in English for specific purposes*. Houndmills, UK: Palgrave Macmillan.

Basturkmen, H. (2012). A genre-based investigation of discussion sections of research articles in dentistry and disciplinary variation. *Journal of English for Academic Purposes, 11*, 134–144.

Bawarshi, A.S., & Reiff, M.J. (2010). *Genre: An introduction to history, theory, research, and pedagogy*. West Lafayette, IN: Parlor Press.

Belcher, D. (Ed.). (2009). *English for specific purposes in theory and practice*. Ann Arbor, MI: University of Michigan Press.

Belcher, D., Johns, A.M., & Paltridge, B. (Eds.). (2011). *New directions in English for specific purposes research*. Ann Arbor, MI: University of Michigan Press.

Bhatia, V.K. (2004). *Worlds of written discourse: A genre-based view*. New York, NY: Continuum.

Biber, D., Connor, U., & Upton, T.A. (2007). *Discourse on the move: Using corpus analysis to describe discourse structure*. Amsterdam: John Benjamins.

Biber, D., Connor, U., Upton, T.A., & Kanoksilapatham, B. (2007). Introduction to move analysis. In D. Biber, U. Connor, & T. A. Upton (Eds.), *Discourse on the move: Using corpus analysis to describe discourse structure* (pp. 23–41). Amsterdam: John Benjamins.

Biber, D., Conrad, S., & Reppen, R. (1998). *Corpus linguistics: Investigating structure and use*. Cambridge: Cambridge University Press.

Bowker, L., & Pearson, J. (2002). *Working with specialized language: A practical guide to using corpora*. New York, NY: Routledge.

Brown, J.D. (1995). *The elements of language curriculum: A systematic approach to program development*. Boston, MA: Heinle & Heinle.

Bruce, I. (2009). Results sections in sociology and organic chemistry articles: A genre analysis. *English for Specific Purposes, 28*, 105–124.

Cheng, A. (2006). Understanding learners and learning in ESP genre-based writing instruction. *English for Specific Purposes, 25*, 76–89.

Douglas, D. (2000). *Assessing languages for specific purposes*. Cambridge: Cambridge University Press.

Dudley-Evans, T., & St John, M.J. (1998). *Developments in English for specific purposes: A multi-disciplinary approach*. Cambridge: Cambridge University Press.

Ferris, D.R. (2005). Reflections of a 'blue collar linguist': Analysis of written discourse, classroom research, and EAP pedagogy. In P. Bruthiaux, D. Atkinson, W.G. Eggington, W. Grabe, & V. Ramanathan (Eds.), *Directions in applied linguistics* (pp. 91–100). Clevedon, UK: Multilingual Matters.

Ferris, D.R. (2011). *Treatment of error in second language student writing* (2nd ed.). Ann Arbor, MI: University of Michigan Press.

Ferris, D.R., & Hedgcock, J.S. (2013). *Teaching L2 composition: Purpose, process, and practice* (3rd ed.). New York, NY: Routledge.

Grabe, W., & Kaplan, R.B. (1996). *Theory and practice of writing: An applied linguistic perspective.* New York, NY: Longman.

Grabe, W., & Zhang, C. (2013). Second language reading–writing relations. In A.S. Horning & E.W. Kraemer (Eds.), *Reconnecting reading and writing* (pp. 108–133). Anderson, SC: Parlor Press/WAC Clearinghouse.

Gunnarsson, B.-L. (2009). *Professional discourse.* London: Continuum.

Hirvela, A. (2004). *Connecting reading and writing.* Ann Arbor, MI: University of Michigan Press.

Hunston, S. (2002). *Corpora in applied linguistics.* Cambridge: Cambridge University Press.

Hyland, K. (1998). *Hedging in scientific research articles.* Amsterdam: John Benjamins.

Hyland, K. (2004a). *Genre and second language writing.* Ann Arbor, MI: University of Michigan Press.

Hyland, K. (2004b). *Disciplinary discourses: Social interactions in academic writing.* Ann Arbor, MI: University of Michigan Press.

Hyland, K., & Bondi, M. (Eds.). (2006). *Academic discourse across disciplines.* Frankfurt: Peter Lang.

Johns, A.M. (1997). *Text, role, and context: Developing academic literacies.* Cambridge: Cambridge University Press.

Johns, A.M. (2007). Genre awareness for the novice academic student: An on-going quest. *Language Teaching, 41*, 237–252.

Kanoksilapatham, B. (2005). Rhetorical structure of biochemistry research articles. *English for Specific Purposes, 24*, 269–292.

Kanoksilapatham, B. (2007). Rhetorical moves in biochemistry research articles. In D. Biber, U. Connor, & T. Upton (Eds.), *Discourse on the move: Using corpus analysis to describe discourse structure* (pp. 73–119). Amsterdam: John Benjamins.

Lin, L., & Evans, S. (2012). Structural patterns in empirical research articles: A cross-disciplinary study. *English for Specific Purposes, 3*, 150–160.

Nation, I.S.P., & Macalister, J. (2010). *Language curriculum design.* New York, NY: Routledge.

National Commission on Writing in America's Schools and Colleges. (2003). *The neglected "R": The need for a writing revolution.* Retrieved July 10, 2012, from ⟨http://www.collegeboard.com/prod_downloads/writingcom/neglectedr.pdf⟩

Nesi, H., & Gardner, S. (2012). *Genres across the disciplines: Student writing in higher education.* Cambridge: Cambridge University Press.

Nwogu, K.N. (1997). The medical research paper: Structure and functions. *English for Specific Purposes, 16*, 119–138.

O'Keeffe, A., McCarthy, M., & Carter, R. (2007). *From corpus to classroom: Language use and language teaching.* Cambridge: Cambridge University Press.

Paltridge, B. (2001). *Genre and the language learning classroom.* Ann Arbor, MI: University of Michigan Press.

Paltridge, B. (2006). *Discourse analysis: An introduction.* New York, NY: Continuum.

Paltridge, B., Harbon, L., Hirsch, D., Shen, H., Stevenson, M., Phakiti, A., & Woodrow, L. (2009). *Teaching academic writing: An introduction for teachers of second language writers.* Ann Arbor, MI: University of Michigan Press.

Peacock, M. (2002). Communicative moves in the discussion section of research articles. *System, 30*, 479–497.

Plakans, L. (2009). The role of reading strategies in integrated L2 writing tasks. *Journal of English for Academic Purposes, 8*, 252–266.

Posteguillo, S. (1999). The schematic structure of computer science research articles. *English for Specific Purposes, 18*, 139–160.
Reppen, R. (2010). *Using corpora in the language classroom*. Cambridge: Cambridge University Press.
Richards, J.C. (2001). *Curriculum development in language teaching*. Cambridge: Cambridge University Press.
Robinson, M.S., & Stoller, F.L. (2007). A read-analyze-write approach to research-related literacy skills for upper-division chemistry majors. In K. K. Katukstis & T. Elgren (Eds.), *Designing, implementing, and sustaining a research-supportive undergraduate curriculum: A compendium of successful curricular practices from faculty and institutions engaged in undergraduate research* (pp. 175–190). Washington, DC: Council on Undergraduate Research.
Robinson, M.S., Stoller, F.L., Costanza-Robinson, M.S., & Jones, J.K. (2008). *Write like a chemist: A guide and resource*. Oxford: Oxford University Press.
Robinson, M.S., Stoller, F.L., & Jones, J.K. (2008). Using the ACS Journals Search to validate assumptions about writing in chemistry and improve chemistry writing instruction. *Journal of Chemical Education, 85*, 650–654.
Samraj, B. (2002). Introductions in research articles: Variations across disciplines. *English for Specific Purposes, 21*, 1–17.
Stoller, F.L., Horn, B., Grabe, W., & Robinson, M.S. (2005). Creating and validating assessment instruments for a discipline-specific writing course: An interdisciplinary approach. *Journal of Applied Linguistics, 2*(1), 75–104.
Stoller, F.L., Horn, B., Grabe, W., & Robinson, M.S. (2006). Evaluative review in materials development. *Journal of English for Academic Purposes, 5*, 174–192.
Stoller, F.L., & Robinson, M.S. (2013). Chemistry journal articles: An interdisciplinary approach to move analysis with pedagogical aims. *English for Specific Purposes*, 32, 45–57.
Swales, J.M. (1990). *Genre analysis: English in academic and research settings*. Cambridge: Cambridge University Press.
Swales, J.M. (1998). *Other floors, other voices: A textography of a small university building*. Mahwah, NJ: Lawrence Erlbaum Associates.
Swales, J.M. (2004). *Research genres: Exploration and applications*. Cambridge: Cambridge University Press.
Swales, J.M., & Feak, C.B. (2011). *Navigating academia: Writing supporting genres*. Ann Arbor, MI: University of Michigan Press.
Swales, J.M., & Feak, C.B. (2012). *Academic writing for graduate students: Essential tasks and skills* (3rd ed.). Ann Arbor, MI: University of Michigan Press.
Tardy, C.M. (2009). *Building genre knowledge*. West Lafayette, IN: Parlor Press.
Tribble, C. (2010). A genre-based approach to developing materials for writing. In N. Harwood (Ed.), *English language teaching materials: Theory and practice* (pp. 157–178). Cambridge: Cambridge University Press.
Wardle, E.A. (2004). Can cross-disciplinary links help us teach "academic discourse" in FYC? *Across the Disciplines: Interdisciplinary Perspectives on Language, Learning, and Academic Writing, 1*. Retrieved June 16, 2012, from ⟨http://wac.colostate.edu/atd/articles/wardle2004/⟩
Williams, I.A. (1999). Results sections of medical research articles: Analysis of rhetorical categories for pedagogical purposes. *English for Specific Purposes, 18*, 347–366.
Willis, D., & Willis, J. (2007). *Doing task-based teaching*. Oxford: Oxford University Press.
Wingate, U. (2012). Using academic literacies and genre-based models for academic writing instruction: A 'literacy' journey. *Journal of English for Academic Purposes, 11*, 26–37.

CHAPTER 3

A genre analysis of student microbiology laboratory notebooks

David I. Hanauer
Indiana University of Pennsylvania

This chapter reports on a qualitative analysis of student laboratory notebooks in a professional microbiology laboratory. The study explored the genre characteristics of laboratory notebooks and organized these characteristics according to social functions, models of notebook completion, and types of entry utilized. To conduct this study, a system for analyzing notebook entry type was modified from Ruiz-Primo, Li, and Shavelson (2002). The results suggest that the laboratory notebook is a multifaceted literacy object that functions as a scientific document for recording laboratory activities, a learning tool for acquiring laboratory protocols, and a personal journal for enhancing scientific engagement.

The laboratory notebook

When investigating scientific literacy, applied linguists and rhetoricians of science have tended to prioritize the research article as the predominant genre of science writing and, accordingly, the majority of studies have been directed towards the exploration and explication of that form of writing (Curry 2014; Gross 1990; Gross, Harmon, & Reidy 2002; Hanauer & Englander 2011, 2013; Hyland 2000, 2005; Hyland & Salager-Meyer 2008; Knorr-Cetina 1981; Salager-Meyer, Defives, & Hamelinsck 1996; Swales 1990). Although the research article may be the most salient and externally accessible form of scientific writing, it is arguably not the most prevalent use of writing in science (Hanauer, Hatfull, & Jacobs-Sera 2009; Latour & Woolgar 1986). Rather, the laboratory notebook is a more ubiquitous and functionally significant form of scientific writing. As documented by various ethnographies of laboratory work, recording and documenting the actions and outcomes of scientific activity as a form of multimodal literacy is a continual process (Hanauer et al. 2009; Latour & Woolgar 1986; Myers 1990). What these studies show is that a far larger amount of scientific information is recorded in notebook form than is ultimately published in research articles. Laboratory notebooks are a crucial component of many scientists' work, ideally providing "a record of everything a scientist does" so that others can "reproduce and validate" results (Sarini, Blanzieri, Giorgini, & Moser 2004, p. 132).

Bearing in mind the significance and centrality of the laboratory notebook for professional scientific work, it is not surprising to find interest in the role of the laboratory notebook as an educational tool in the training of students and novice scientists. For example, Hanauer et al. (2009) argue that the laboratory notebook has "enormous potential" as an educational tool for teaching and assessing scientific inquiry because "it provides an in-depth, up-to-date description of the procedures, outcomes and thoughts concerning the process of scientific inquiry conducted by the student" (p. 103). Likewise, Ruiz-Primo (1998) describes the laboratory notebook in educational contexts as "a compilation of entries that provide a partial record of the instructional experiences a student had in her or his classroom" (p. 4). A core assumption behind the educational use of laboratory notebooks is that they reflect authentic practices of scientists and as such may be valuable in educational science processes as well (Hargrove & Nesbit 2003).

The aim of this chapter is to explore the use of laboratory notebooks written by students as novice scientists in an educational program designed to teach and facilitate the experience of professional scientific inquiry in a microbiology laboratory. The methodological approach taken in this study is qualitative, utilizing textual analysis, observation, and interview methods. This chapter contributes a genre analysis which aims to explicate the social functions, procedural components, and structural features of laboratory notebooks (Hanauer 1998, 2004, 2006).

Educational studies of laboratory notebooks

Within science education there is increasing emphasis on providing laboratory-based research experiences for students (Desai, Gatson, Stiles, Stewart, Laine, & Quick 2008; Wood 2003). The educational concept which underpins this approach is based on the importance assigned to inquiry experiences in facilitating the development of the types of thinking that scientists engage in. As defined by the National Science Foundation (2000), inquiry learning is "an approach to learning that involves a process of exploring the natural or material world, and leads to asking questions, making discoveries, and rigorously testing these discoveries in the search for new understanding" (p. 1). The idea behind the increased usage of laboratory courses, simply put, is that exposure to and experience with real science in professional settings is the best method for acquainting students with the procedures, social settings, outcomes, and values of scientists (Hanauer et al. 2009). This apprenticeship model focusing on inquiry has a history that can be traced to the work of Dewey (1938) and Bruner (1961).

In inquiry learning research courses, the laboratory notebook is commonly used by students to record their research activities. In principle, a laboratory notebook should include a "question to explore, predictions, a description of what was done, and what students learned," including "narrative statements and drawings about the student's observations, data sets, diagrams, graphs and tables" (Hargrove & Nesbit 2003, p. 2). This conception of the laboratory notebook sees it as a central part of the process that a student undergoes in learning the scientific practices of the laboratory. But Ruiz-Primo et al. (2002) suggest that students do not fully understand the components of a laboratory notebook and fail to report complete experiments, as they are usually missing "the purpose/objective of the experiment or data interpretation and conclusions" (p. 12).

Bearing in mind the educational directive of large funding bodies such as the National Science Foundation, the Howard Hughes Medical Institute, and the National Institutes of Health to increase the frequency of research experiences for students in science and the centrality of laboratory notebooks for this process, understanding the literacy components of the laboratory notebook is important. The current chapter aims to address this need and use applied linguistics methods to investigate the laboratory notebook. Specifically, the research question this study aims to answer is:

What are the genre characteristics of the student laboratory notebook?

Methodology

Research site

This study was conducted in a professional microbiology laboratory at a major university in Pennsylvania known for its integration of various student populations into authentic scientific inquiry. This laboratory is unusual in that it serves several types of student-researchers with very different educational backgrounds. The laboratory includes postdoctoral researchers, doctoral students, and undergraduates who interact in the core research agenda of the laboratory. The laboratory is known for its educational initiative which brings students into the laboratory to work with professional researchers on a shared research program.

Data collection

Data collection involved observation, document collection, and interviews. The project was conducted in these stages:

1. *Observation of Laboratory Notebook Use*: To explore the processes of production of the notebooks, student use of notebooks during ongoing laboratory work was observed. Observations were recorded using field notes. Over a two-week period, observations were conducted (and field notes made) that consisted of watching the ways that nine participants used the notebooks. Special attention was directed at how notebook writing was integrated in the processes of conducting laboratory research. The observations were conducted by the author, who is well known to all members of the laboratory.
2. *Participant Interviews*: Following the period of observation, interviews about the process of writing notebooks were conducted with six members of the laboratory: two graduate students and four undergraduates. During the interviews, notes were taken and fully written up following the interview. Using observation and interview notes, a description of the processes of production for the notebooks and their social function was developed. These descriptions of the processes of production of laboratory notebooks and their social functions were validated and modified through follow-up discussions with participants.
3. *Structural Analysis of Laboratory Notebooks*: For the current study, 51 completed notebooks of finalized inquiry projects for a two-year period were collected for analysis.

Data analysis

Field notes were analyzed to develop understandings of the social functions of the laboratory notebooks. Explicit statements that participants made about their laboratory notebooks as they worked in the laboratory were used to construct an understanding of notebooks' social functions. Specifically, these statements were analyzed for propositions that positioned the notebook usage. The resulting analytic description was then presented to participants to obtain their perspectives on the conclusions, an aspect of qualitative research known as member checking (Giacomini & Cook 2000). To understand the inquiry learning and psychological processes of notebook writing, the focus of analysis was the chronological development of the actions recorded in the field notes and concurrent digital photography. Using a notebook was defined as involving a series of recurring actions. The result of this analysis was a description of the ways that students used their laboratory notebooks.

To analyze the structural and informational features of laboratory notebooks, a categorization system was developed for coding the data. Developing this system entailed a cyclical process of reading, categorizing, defining, and the subsequent coding of the notebooks. To develop a coding system, the author adapted the ten-category system developed by Ruiz-Primo et al. (2002), which is presented below. Initially this system was applied to the coding and categorization of specific entries in 20% of the sample of 51 laboratory notebooks (10 notebooks) to test the validity of the coding system. This process involved applying the initial coding system to notebook entries, redefining these coding categories if necessary, adding new categories, and eliminating others in a recursive model. Each coded element was defined and examples extracted. The modified version of Ruiz-Primo et al.'s (2002) coding system comprehensively covers the types of entries made in the notebooks. The final coding system consists of these 10 categories:

1. *Defining/Describing* (*Verbal*): A 'defining or describing verbal' entry describes a term, concept, or procedure to be performed by the writer. The aim of this type of entry is educational; it allows the notebook writer to have access to new substantive and procedural microbiological knowledge. An entry of this type may involve, for example, writing out a full protocol for a particular microbiological procedure.
2. *Defining/Describing* (*Pictorial*): A 'defining or describing pictorial' entry provides a visual or graphic representation of a procedure, research tool, or microbiological concept (e.g. an organism) that relates to the writer's scientific inquiry. As with the verbal description, the aim of this type of entry is educational and allows the writer to understand how particular research tools or results can be understood and used. An entry of this type may also involve the drawing of a schematic version of the tool.
3. *Designing an Experiment*: A 'designing an experiment' entry provides a description of a future experiment to be conducted. The design is the basic template of the scientific investigation to be conducted. It does not refer to protocol instructions, but rather to the design of a contextualized study that the researcher will conduct and the reasons for this design.
4. *Reporting a Problem*: A 'reporting a problem' entry specifies a problem that needs to be solved, which is usually connected to the outcome of an experiment or procedure that did not go as expected.
5. *Verbal Reporting of a Performed Procedure*: The 'verbal reporting of a performed procedure' entry reports in writing on a procedure that has been conducted. In its short form, the procedure is merely named. In such cases, the protocol may be referenced in another part of the notebook or in a laboratory manual. In its full form, a detailed description of the protocol that has been followed is written in the notebook.

6. *Visual Reporting of a Performed Procedure*: The 'visual reporting of a performed procedure' entry presents a pictorial representation of the procedure performed. This type of entry is particularly useful if the procedure involved the use of positional information such as the placement of particular restriction enzymes in a gel. The pictorial representation is usually schematic.
7. *Verbal Reporting of Results*: A 'verbal reporting of results' entry presents a verbal or mathematical description of the results that a specific procedure or experiment achieved. These results can be detailed or consist of only a limited specification of the highlights or points of interest in the results.
8. *Visual Reporting of Results:* A 'visual reporting of results' entry provides visual evidence of an outcome of a procedure or experiment, usually a digital photograph of a gel or a dish. However, other procedures such as electromicroscopy may also produce a visual representation. The visual reporting of results entry consists of the attachment of the picture, or in some cases, the hand drawing of the result, to the notebook.
9. *Interpreting Data and Reaching Conclusions:* An 'interpreting data and reaching conclusions' entry presents the researcher's understandings of the results of a procedure or experiment, which might be partial.
10. *Personal Statement*: A 'personal statement' entry presents a comment that is not directly related to the scientific inquiry process. This may include social comments about meetings with other students, details of local restaurants, phone numbers, and meetings of various sorts. There may also be emotive meta-comments concerning the work conducted by the researcher (such as an expression of excitement and joy as a result of naming a discovered bacteriophage virus).
11. *Mentor Comments:* A 'mentor comments' entry is written by one of the student's laboratory mentors in relation to the work conducted by the student, following the mentor's review of the notebook.
12. *Meeting Reports*: A 'meeting report' entry documents the writer's attendance at a laboratory meeting and/or contents of the meeting.

Using this coding system, the whole data set was coded. To ensure that the coding of the notebooks was reliable, each notebook was coded twice by the researcher at two different times approximately one month apart, the second taking place without consulting the initial coding (Weber 1990). Rather than having two coders for inter-rater reliability, this intra-rater approach was employed as a result of the disciplinary specific nature of the notebooks and the lack of other applied linguistics or educational research expertise in the laboratory. The data from the two coding sheets for each notebook were compared. Intra-rating agreement was calculated at 87%. Following the coding of each notebook, a table was compiled of the average percentage of usage of each code category.

Results

The results section is organized with sections describing each of the laboratory notebook genre components of social functions, processes of production, and structural features.

Social functions

On the inside cover of the notebooks was affixed a label created by the notebook manufacturer, which relates the social role of the laboratory notebook. In this inscription, students are informed that the notebook is "an essential central record of what you have done in the lab" and that "all procedures and results must be clearly described." The same inscription goes on to specify that the "lab notebook should be written as a document that can be read by others so that experiments can be easily repeated." According to the manufacturer, the social function of the notebook is as a recording device that will allow researchers easily to follow the scientific procedures that were performed and to see the results found.

Data from participant interviews concurred with the notebook manufacturer's assumption that the role of the notebook is as a recording device that allows specific procedures, quantities, and outcomes to be recorded over time. As noted by one graduate student participant, "It allows me to look at an experiment I did last year and state exactly what I did." All participants interviewed agreed that the role of the notebook was to allow other researchers to ascertain what was done and when, so that research can be repeated and results properly understood. Individual respondents also provided descriptions of additional roles for the notebook. One graduate student participant specified that the notebook performed the role of a "diary that recorded the evolution of ideas" during inquiry. An undergraduate participant specified a similar concept in that the notebook had the function of helping her "creatively problem solve" various issues. Another undergraduate participant specified that the lab notebook helped her to organize her research studies.

In interview statements made by members of the laboratory, there is extensive agreement about the social function of the laboratory notebook as an important scientific document that should record all research activity conducted by the student. They believe it should be written in a format that would allow another researcher to be to able follow what was done. The function of the notebook is to provide a clear, comprehensive, and detailed record of the procedures and outcomes of the scientific inquiry process conducted by the researcher.

Processes of production

The production of the laboratory notebook is an extensive process that covers weeks, months, or years of activity. The notebook is used on a daily basis – and throughout the day – as a record of activities conducted as part of scientific inquiry. Completing a notebook is an extended writing process conducted in stages over a long period of time. Through interviews and observations within the laboratory, two models of daily writing with notebooks were defined: *Interactive Notebook Development* and *Post-procedure Notebook Development*. To a certain extent, any given notebook might have a mixture of both types of process. The difference between the two types of process consists in the order of recording information in the notebook. In the *Interactive Notebook Development* model, protocols for future actions in the laboratory are recorded. This model functions as a learning tool in which the writer records novel protocols in order to understand them properly so they can then be used to conduct scientific inquiry in the laboratory. In the *Post-procedure Notebook Development* model, the protocols of a known scientific experiment or process are recorded after the actions have taken place in the laboratory; these protocols are usually referred to in an abbreviated manner (sometimes referencing the pages of a laboratory protocol manual).

Based on interview and observational data, the interactive process tends to precede the post-procedure process and is characteristic of a learning stage of activity in the laboratory. As work in the laboratory begins with a student learning procedures and protocols and repeating them many times thereafter, the writing process is often initially interactive but later becomes post-procedure once the student-writer-scientist has internalized the protocol and can function without referring to the notebook. In the following subsections each of these writing-learning processes will be explicated.

Interactive notebook development

The interactive notebook development model may be involved in the case of new researchers in the laboratory or seasoned researchers embarking on new experiments and procedures. In the former case, novice researchers working within the apprenticeship model of instruction may have little knowledge of laboratory procedures of the thinking processes of a scientist. The writing out of the procedures from reference books or instructions from others in the laboratory is a learning process for the student; protocols are used directly as a reference while conducting the procedures. In the latter case, the interactive use of notebook writing process constitutes the development of new knowledge as procedures may become well-established protocols over time. For the novice, the writing in the notebook becomes a form of problem solving in which specific experiments that may allow the resolution of specific problems can be addressed.

Characteristic of the interactive process is the role of notebook writing early in the novice's experience. The potential procedure to be conducted is researched and written in the notebook before any actual work in the laboratory is conducted. The protocol may be copied from a manual or written from a description provided by a more experienced member of the laboratory. During the scientific inquiry, the notebook tends to be opened to the page on which the protocol has been written out and constantly referred to during the laboratory work. As results appear (whether from a short- or long-term experiment), they are recorded in the notebook. If the results have a visual component, they are documented by printing a digital picture or computer output and including it in the notebook. Sometimes a written description of the results is added to the notebook but with differences in the degree of detail provided. In the interactive process, the notebook is an integral part of laboratory work and is referenced many times. Results and comments are recorded before, during, and after the laboratory work as represented in Figure 3.1.

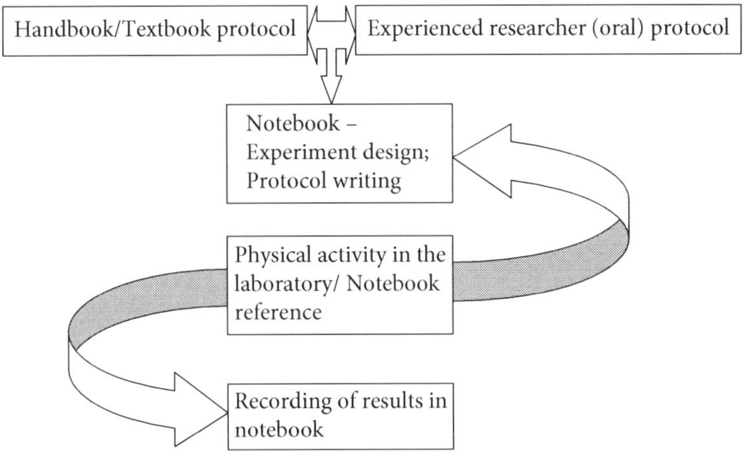

Figure 3.1. Schematic representation of interactive notebook writing development

Post-procedure notebook development

Post-procedure notebook development is used when a student-writer-scientist has more experience with and knowledge of a set of commonly used laboratory protocols and procedures. As with any laboratory context, certain procedures occur highly frequently and, as the student researcher gains confidence in using these procedures, the notebook is less frequently consulted. The post-procedure aspect of notebook usage is that the laboratory work is performed and completed, including the visual recording or results, before the procedure is written in the notebook. It is only at the point of having results that the notebook is used. In writing the procedure in the notebook, abbreviated forms of the procedure are often used, such as a brief description of the procedure (without detail), reference to the name of the procedure alone, or reference to another source or page of the notebook on which the protocol is fully explicated. The results of the procedure or experiment are presented in visual format by printing and then pasting into the notebook a digital picture or print out. There is no detailed description of the results, but rather a brief specification of the major finding and some evaluative comments on what worked and what did not. The post-procedure process has three basic components: the to-do list, physical laboratory activities, and the scientific notebook. These are defined as:

1. The *To-do List*: A to-do list that has been generated the day before (or several days before) is reviewed. This list may be on the computer in the form of a sticky notes program, placed on real sticky notes, or written on a notepad and not in the notebook itself. The to-do list is usually not retained but is discarded after the procedure is finished.
2. *Physical Scientific Activities*: The experiments or procedures of the day are conducted in the laboratory. The notebook may be referred to if the researcher needs to consult any protocols recorded in it. However, for some participants, the protocols exist in other formats (computer files, binders, books, etc.), not in the notebook. Accordingly, the notebook may not be referred to at all.
3. *The Scientific Notebook*: Following the completion of the laboratory work, a written record is kept in the notebook. This record provides differing degrees of information about the laboratory work conducted. If the procedure is well known to the researcher, it is not written out but rather referred to by the source of the original protocol (the page number in the notebook where the protocol is written; the binder in which the protocol appears, etc.). If the procedure takes place frequently, it may be referred to by name and just stated that it was conducted. The procedure performed is noted when the results have been produced, which often involves including visual data in the form of a digital picture of a gel or a dish. The visual results are not verbally explained but special features are often pointed out with evaluative comments. Problems may be noted as well. A representation of this process is in Figure 3.2.

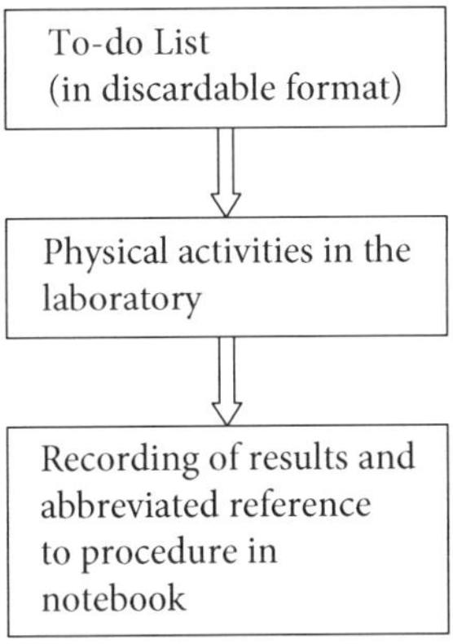

Figure 3.2. Schematic representation of the post-procedure process of notebook development

The two ways of entering items into the laboratory notebook differ in relation to the role of the notebook within the actual processes of conducting laboratory research. They seem to be related to the degree of acquaintance and confidence the student-researcher has in connection to the procedure being conducted. In the observational and interview data collected for this study, an interactive writing process in which the notebook was addressed during the laboratory work was characteristic of a learning stage of laboratory work. Once protocols had been learnt and student researchers were more experienced, a post-procedure approach was utilized for notebook completion. Since new protocols are learnt throughout one's education as a laboratory researcher, student-writer-scientists move between the two models of notebook writing at different stages of their education. It should be noted that in principle, when a post-procedure writing process is used, the writer replaces an actual record of what was done in the laboratory work with a standardized description of the same procedure based on the accepted protocol. Although these might indeed be the same, an accurate continual description of laboratory work (as required in relation to the social function of the laboratory notebook) would seem to necessitate the usage of an interactive writing process and not the post-procedure approach.

Structural features of scientific notebooks

This analysis involved a modified form of Ruiz-Primo et al.'s (2002) coding system. To be able to compare and average the usage of the different categories of the coding system, the raw frequency of entry by category was transformed into a percentage of the total number of entries in each notebook. These percentages were then averaged across the notebooks. Table 3.1 presents the average frequency percentage of usage by category for the notebooks in the data set.

Table 3.1. Percentage of category usage across laboratory notebooks

Category	Frequency (percentage)
Defining/Describing – Verbal	4.76%
Defining/Describing – Visual	1.44%
Designing an Experiment	0.57%
Reporting a Problem	2.27%
Reporting Procedure – Verbal	46.05%
Reporting Procedure – Visual	2.39%
Reporting Results – Verbal	10.43%
Reporting Results – Visual	22.02%
Interpreting Data/Reaching Conclusions	1.06%
Personal Statements	7.20%
Mentor Comments	1.20%
Meeting Reports	0.96%

As Table 3.1 shows, the structural features of notebooks are characterized by high frequencies of reporting procedures and reporting results. Together the four categories of verbal and visual reporting of procedure and results comprise 80.89% of entries in the laboratory notebooks. The highest frequency category is that of the verbal reporting of procedures, which constitutes 46.05% of all entries and reflects the predominance of physical laboratory inquiry that is central to the process of bacteriophage identification. While 32.45% of entries dealt with the reporting of results, visual data were more frequently used to report results than were verbal descriptions. Thus, from a frequency perspective, entries in the notebooks are overwhelmingly characterized by verbal description of procedure and visual presentation of results. Taken together, these two categories make up 68.07% of entries.

The results of the analysis of the coded items of the notebooks suggest an education process in which student researchers are learning laboratory protocols and research procedures. There is extensive description of procedures and presentation of results, but there is relatively little for results interpretation (1.06%), reporting of a problem (2.27%), or design of an experiment (0.57%). In this sense, these results are close to the findings of Ruiz-Primo et al. (2002), suggesting that students do not tend to focus on experiment design or interpretation in their laboratory work.

It should also be noted the 7.2% of all notebook entries involved personal comments. This is a relatively high category and suggests a slightly different understanding of what goes into a laboratory notebook. These entries can be understood as a way of personalizing the laboratory notebook, which is particularly important in terms of current interest in the psychosocial variables of science education (Hanauer, Frederick, Fotinakes, & Strobel 2012). Thus statements which express a student's excitement over discovering a novel organism might not be "functional" from the perspective of the expressed social function of laboratory notebooks, but they allow the student to connect and engage with their research and as such may play an important role in enhancing the research experience.

Characterizing the genre of microbiology laboratory notebooks: Conclusions and implications

The aim of this study was to provide a genre analysis of laboratory notebooks. The results suggest that the laboratory notebook is a multifaceted literacy object that functions on several different levels. On one level, the laboratory notebooks are intended to be scientific documents that record laboratory activities. The notebooks do include extensive description of procedures and visual presentation of results. The presence of these types of entry documents the activities conducted by the student in the laboratory and provides laboratory personnel with access to the research conducted. However, as noted in the description of the two processes of writing, when a more interactive writing process is used, this documentation can be more accurate. In contrast, the use of a post-procedure writing process runs the risk of presenting an assumed set of procedures that were followed rather than what was actually done.

On another level, these notebooks are a learning tool. The interactive writing process was characteristic of learning a procedure. The post-procedure writing process was used when the procedure had actually been learnt and the student felt confident enough so as not to need to reference the notebook and involved far fewer entries (since the procedure was only referenced in abbreviated form). Of all notebook entries, 46.05% were written descriptions of laboratory procedures, which suggests that the learning of protocols was a significant component of laboratory work for these student-writer-scientists.

On yet a third level, the notebooks provide evidence of personal engagement with science. Of all notebook entries, 7.2% were of a personal nature. The presence of these items suggests that the student-writer-scientist sees the laboratory notebook as a form of journal in which personal insights and events can be recorded. This finding does not invalidate the laboratory notebook as a scientific document but does suggest a degree of personal involvement with scientific activity in this research experience. Current understandings of science education suggest that personal engagement with research experiences is important for retention of students in the sciences (Hanauer et al. 2012; Hunter et al. 2006). Accordingly, it is possible that comments of a personal nature (especially meta-emotive commentary), while not directly scientific, should be encouraged in laboratory notebooks (or at least tolerated).

Some issues of concern for science educators arise from the analysis of notebooks presented here. The results of this study parallel previous research that suggests that student-researchers do not put enough emphasis on the understanding of the purposes/design of experiments or the interpretation of results. This might be a knowledge threshold issue in that student-researchers may not have enough scientific understanding to be able to really interpret their results or design experiments. However, using notebooks is a central part of any authentic scientific research activity and needs to be modeled in research experiences. Such modeling could include monitoring and discussion of notebook results with more experienced professionals and requesting that student researchers pay attention to laboratory notebook completion.

For science-educators, the current study provides a methodological approach that can be utilized in the investigation of other student research settings. The coding system presented here and modified from Ruiz-Primo et al. (2002) can be applied in other contexts and results compared. As a hypothesis, notebook completion should relate to the type of educational program within which notebooks are used. Accordingly the coding of laboratory notebook entries might be useful in differentiating among various research experiences. Within the science education literature, there is an increasing understanding that not all research experiences are equal and a demand for ways of assessing these differences (Hanauer et al. 2012; Hunter et al. 2006).

Some provisional educational conclusions emerge from this study. The results suggest that it might be beneficial for science educators to interact with student researchers in relation to the design, purpose, problems, and conclusions of the investigations conducted and require that these discussions become entries within the laboratory notebook. Increased mentorship of this type should aim to model the types of thinking that authentic research entails and as such might deepen the student's understanding of research and what science actually involves. The use of a post-procedure writing process should be carefully considered by the laboratory instructors for possible dangers in relation to the accurate description of laboratory work. While descriptions might be artifacts of a student having learnt a procedure, the descriptions might not be an accurate in terms of what was actually done. For educational and scientific purposes, instructors may prefer that students use the interactive process and not transfer to the post-procedural approach even if they know the procedure well. Finally, emotive and personal comments included in the laboratory notebook should be considered as providing information about students' excitement, frustration, and engagement with scientific activities.

Overall, the study conducted here provides insight into the options, processes, and difficulties of using laboratory notebooks as part of research experiences. As demonstrated in this chapter, when used by students in research experiences, the laboratory notebook takes on more roles than just the documentation of laboratory activities. It is a learning tool and has the potential to enhance student engagement. As such, the laboratory notebook has potential to facilitate enhanced understanding of the sciences on both the cognitive and emotive levels.

References

Bruner, J.S. (1961). The act of discovery. *Harvard Educational Review, 31,* 21–32.

Curry, M.J. (2014). Graphics and invention in academic engineers' writing for publication. In M. J. Curry & D. I. Hanauer (Eds.), *Language, literacy, and learning in STEM education: Research methods and perspectives from applied linguistics* (pp. 87–106). Amsterdam: John Benjamins.

Desai, K.V., Gatson, S.N., Stiles, T.W., Stewart, R.H., Laine, G.A., & Quick, C.M. (2008). Integrating research and education at research-extensive universities with research-intensive communities. *Advances in Physiological Education, 32,* 136–141.

Dewey, J. (1938). *Logic: The theory of inquiry.* New York, NY: Henry Holt.

Giacomini, M.K., & Cook, D.J. (2000). Qualitative research in health care: Are the results of the study valid? *Journal of the Americal Medical Association, 284*(3), 357–362.

Gross, A.G. (1990). *The rhetoric of science* (2nd ed.). Cambridge, MA: Harvard University Press.

Gross, A.G., Harmon, J.E., & Reidy, M.S. (2002). *Communicating science: The scientific article from the 17th century to the present.* West Lafayette, IN: Parlor Press.

Hanauer, D.I. (1998). The genre-specific hypothesis of reading: Reading poetry and reading encyclopedic items. *Poetics, 26*(2), 63–80.

Hanauer, D.I. (2004). Silence, voice and erasure: Psychological embodiment in graffiti at the site of Prime Minister Rabin's assassination. *Psychotherapy in the Arts, 31*(1), 29–35.

Hanauer, D.I. (2006). *Scientific discourse: Multiliteracy in the classroom.* London: Continuum.

Hanauer, D.I., & Englander, K. (2011). Quantifying the burden of writing research articles in a second language: Data from Mexican scientists. *Written Communication, 28*(4), 403–416.

Hanauer, D.I., & Englander, K. (2013). *Writing science in a second language.* West Lafayette, IN: Parlor Press.

Hanauer, D.I., Frederick, J., Fotinakes, B., & Strobel, S. (2012). Linguistic analysis of project ownership for undergraduate research experiences. *CBE-Life Sciences Education, 11,* 378–385.

Hanauer, D.I., Hatfull, G.F., & Jacobs-Sera, D. (2009). *Active assessment: Assessing scientific inquiry.* New York, NY: Springer.

Hargrove, T.Y., & Nesbit, C. (2003). *Science notebooks: Tools for increasing achievement across the curriculum* (ERIC Digest ED482720). ERIC Clearinghouse for Science Mathematics and Environmental Education.

Hyland, K. (2000). Hedges, boosters and lexical invisibility: Noticing modifiers in academic texts. *Language Awareness*, *9*(4), 179–197.

Hyland, K. (2005). Stance and engagement: A model of interaction in academic discourse. *Discourse Studies*, *7*(2), 173–191.

Hyland, K., & Salager-Meyer, F. (2008). Scientific writing. *Annual Review of Information Science and Technology*, *42*(1), 297–338.

Hunter, A.B., Laursen, S.L., & Seymour, E. (2006). Becoming a scientist: The role of undergraduate research in students' cognitive, personal and professional development. *Science Education*, *91*(1), 36–74.

Knorr-Cetina, K. (1981). *The manufacture of knowledge: An essay on the constructivist and contextual nature of science*. Oxford: Pergamon Press.

Latour, B., & Woolgar, S. (1986). *Laboratory life: The construction of scientific facts*. Princeton, NJ: Princeton University Press.

Myers, G. (1990). *Writing biology*. Madison, WI: University of Wisconsin Press.

National Science Foundation. (2000). *Foundations: Inquiry, thoughts, views and strategies for the K-5 classroom*. Arlington, VA: National Science Foundation.

Ruiz-Primo, M.A. (1998). *On the use of students' science journals as an assessment tool: A scoring approach*. Unpublished manuscript, Stanford University, School of Education.

Ruiz-Primo, M.A., Li, M., & Shavelson, R.J. (2002). *Looking into students' science notebooks: What do teachers do with them*? (CSE Technical Report 562). Los Angeles: Center for the Study of Evaluation, National Center for Research on Evaluation, Standards, and Student Testing, University of California, Los Angeles.

Salager-Meyer, F., Defives, G., & Hamelinsck, M. (1996). Epistemic modality in 19th and 20th century medical English written discourse: A principal component analysis. *Interface. Journal of Applied Linguistics*, *11*(2), 95–117.

Sarini, M., Blanzieri, E., Giogini, P., & Moser, C. (2004). From actions to suggestions: Supporting the work of biologists through laboratory notebooks. In F. Darses, R. Dieng, C. Simone, & M. Zackland (Eds.), *Cooperative systems design: Scenario-based design of collaborative systems* (pp. 131–146). Amsterdam: IOS Press.

Swales, J.M. (1990). *Genre analysis: English in academic and research settings*. Cambridge: Cambridge University Press.

Weber, R.P. (1990). *Basic content analysis*. Newbury Park, CA: Sage.

Wood, W.B. (2003). Inquiry-based undergraduate teaching in the life sciences at large research universities: A perspective on the Boyer Commission Report. *Cell Biology Education*, *2*, 112–116.

CHAPTER 4

Writing and becoming a scientist

A longitudinal qualitative study of three science undergraduates

Neal Lerner and Mya Poe
Northeastern University

In many communication-intensive (CI) classes in the sciences, students are learning the essential practices for knowledge production in tandem with professional communication practices. In short, they are learning to communicate as a biologist or chemist or physicist. To explore this process more closely, we conducted a three-year qualitative study of 17 undergraduate and graduate students in science and engineering CI classes at the Massachusetts Institute of Technology via initial and end-of-year surveys, one-to-one interviews, classroom observation, and collection and analysis of student writing. In this chapter, we present the findings about three students who were sophomores at the beginning of the study and graduating seniors at its conclusion to understand if these students evidenced changes in their understanding of the roles and purposes of communicating science. Based on coded interview data and analysis of student writing, the case studies we report in this chapter offer insights into the intertwined relationship between students' developing scientific knowledge and their shifting identities as students/scientists/engineers/writers. Understanding students' perspectives on entering professional communities through their communicative practices, particularly given the constraints of school contexts, speaks to the need to design instruction that responds to students' developmental levels and their readiness to learn to communicate in science and engineering.

Introduction

An increasingly common curricular requirement at many American colleges and universities are writing-intensive (WI) or communication-intensive (CI) courses, often part of larger writing across the curriculum (WAC) or writing in the disciplines (WID) efforts. While the shape and specifics of these courses vary considerably (Thaiss & Porter 2010), one shared feature is that they embed writing instruction and practice within the existing curriculum. In other words, rather than taking an add-on writing component or a completely separate course, in CI classes in the sciences students are learning the practices essential for knowledge production and communication in those fields while learning technical content. In short, they are studying what it means to communicate as a biologist or chemist or physicist. Put another way, learning to communicate becomes a means *for* learning that technical content. In best-practice CI classrooms, this learning is enabled by writing, speaking, and visualizing tasks that are *authentic* or are modeled on the kinds of communication activities in which professionals in those fields regularly participate (Freedman & Adam 2000). As a consequence, teaching and learning to communicate in these disciplines is modeled on the processes by which novices learn expert practices (Carroll 2002; Haswell 1991; Herrington & Curtis 2000; Sommers & Saltz 2004; see Russell 2001, for a review of naturalistic studies in WAC/WID).

Such curricular efforts offer research opportunities to explore how well this ideal of learning maps onto the reality of our sites of literacy instruction, whether classroom, laboratory, writing center or professional practice site. Research on student learning in these contexts – and particularly for students in STEM disciplines – has demonstrated that student writing development is not a straightforward trajectory from novice to professional (e.g. Artemeva 2005; Beaufort 2007; Blakeslee 2001; Haas 1994; Patton 2011). This body of research consistently points to the ways that academic contexts have complicating factors, limiting the ways that students transfer the communication they do in courses to what they might do as professionals (Artemeva 2009; Dannels 2000; Freedman & Adam 2000). They further point to the longitudinal nature of student writing development: Students' understanding of epistemological practices develops over the time they spend in school and continues developing after graduation and into the workplace (Beaufort 2007; Leydens 2008).

To explore the ways that STEM students learn to communicate, in a previous work we conducted a qualitative study of 17 students, both undergraduate and graduate, in science and engineering CI classes at the Massachusetts Institute of Technology (MIT; Poe, Lerner, & Craig 2010). While our research was motivated in part by concerns about the efficacy of the WAC program in which we were instructors and administrators, as literacy researchers with longstanding interest in student development and scientific knowledge making, we realized the research potential for what we were attempting, particularly in light of the relative lack of research on students' development of rhetorical understanding in STEM disciplines (Beaufort 2007; Ellis 2004). More specifically, we wanted to explore if students would eventually see scientific communication as a persuasive act with necessary attention to audience, purpose, context, and content.

Over the course of one academic year, we followed our research participants as they completed writing tasks in CI courses, and we tracked changes in students' understanding about the purpose and processes of scientific writing and speaking. (We describe this methodology in the next section.) We found that in early CI classes – classes taken in the first or second year of college – most students saw communicating science as merely the translation of sound scientific findings rather than a rhetorical/persuasive act situated in particular contexts. More specifically, students often viewed their communicating science as indexed to school-based contexts, that is, writing activities completed to receive a grade. Upper-level students and graduate students in CI courses, on the other hand, saw communication as highly rhetorical, largely a matter of persuading a reader that the science was valid, interpretations sound, and methods reliable – although there was much variability in students' ability to produce texts that successfully produced those effects. We also found that the nature of school-based tasks precluded how well students could project the writing they did into their professional futures. In classes where students wrote to satisfy their instructor's needs and values, they were unsure how universal those needs and values might be. In other courses, students who showed an emerging professional identity could better project how writing might play a role in their future professional lives, such as in writing a grant proposal to the National Institutes of Health.

Two years later, we returned to eight students from our original study to learn more about changes in their writing development in and beyond CI courses. In this chapter we report on the three sophomores from the 2008 study, who were all about to graduate in 2010. We wanted to understand if the trajectory that we previously mapped across different groups was consistent for these same students across time. Did these sophomore-now-senior students evidence the changes in their understanding of the roles and purposes of writing that we had found previously with upper-level students?

As we report in this chapter, our follow-up study reinforced many of the findings from the initial study, but we also learned about the intertwined relationship between students' developing scientific knowledge and their shifting identities as students/scientists/engineers/writers. More specifically, each of the three case study students had changed majors between sophomore and senior year, a process of identity formation and reformation not uncommon to undergraduates in the United States. However, with these students, that identity shift was closely tied to shifts in their understanding of what it meant to communicate in their previous and new fields. Their success with that communication played a strong role in their decisions to change majors and futures. In a sense, then, students were establishing their "discursive identities" (Brown, Reveles, & Kelly 2005) through the choices they made about whether to learn certain kinds of disciplinary communication.

In what follows, we describe the methods and context for both phases of our study, describe our findings with respect to our three participants, and offer implications for further research and teaching.

Context and methods

We relied on case-study methodology in an effort to understand the "how" and "why" (Yin 2008) of the students' experiences with learning to communicate in science and engineering. Traditional case study research is appropriate for the questions we were asking and the data to which we had access: How do students' understanding of what it means to communicate science change over time? How is that change represented in interviews, written documents, and classroom interactions? Importantly, case study research does not aim to present the case as generalizable, but, instead, in Yin's words, "case studies, like experiments, are generalizable to theoretical propositions and not to populations or universes" (2008, p. 15).

Site

As noted, this study focuses on three undergraduates' communication experiences in communication-intensive (CI) classes. MIT is a four-year, Ph.D.-granting technical university. All MIT undergraduates are required to take four CI-classes, with two of those classes in the student's major. MIT's Writing Across the Curriculum Program supports many of these majors by providing instructors who teach some aspects of the CI component and offer feedback to technical instructors on assignment design, evaluation of student writing, and communication pedagogy. Students were initially selected from two CI courses: Experimental Molecular Biology (EMB) and Communication and Laboratory Fundamentals of Biological Engineering (BE). Both courses share several attributes: They are the first of two required CI courses in the Departments of Biology and Biological Engineering, respectively. In both courses, students pursue novel research specific to the interests of the faculty teaching the courses; and in both courses, experimental research is focused on understanding the relationship between the structure and function of biological systems at the molecular level. Students in EMB are also enrolled in a stand-alone scientific communication course (SciComm) in which they learn the rhetorical fundamentals of creating and communicating scientific research and ultimately produce a research article based on their laboratory findings. BE, on the other hand, includes imbedded communication instruction: Instructors from the WAC Program offer several workshops during lab time, respond to drafts of students' writing, and give feedback to students' oral presentations, while working alongside technical faculty.

Participants

As noted, the three participating students were sophomores when the study began in 2008 and graduating seniors in 2010. Students were selected on a voluntary basis from two CI courses – EMB/SciComm and BE – in which we were communication instructors, which allowed us access to course materials and other artifacts as well as access to course "insiders" (i.e. faculty members). Although we knew these contexts well, we chose student participants for whom we did not have instructional responsibility, thus mitigating potential ethical conflicts had we been responding to and evaluating students' work. In 2008, two of the students (using the pseudonyms April and Carla) were enrolled in EMB/SciComm, while the third (using the pseudonym Nedra) was in BE (see Table 1). All three women were majoring in historically male-dominated fields.

Table 4.1. Participants' course contexts, original and final majors

Name	Course context (2008)	Course context (2010)	Major in original study (2008)	Major at graduation (2010)
April	EMB/SciComm	Biology project laboratory	Chemical engineering	Biology
Carla	EMB/SciComm	Political science laboratory course	Biology	Political science
Nedra	Biological Engineering	Independent thesis project	Biological engineering	Mechanical engineering

Data collection

In light of our interest in students' scientific knowledge-making practices, the initial step upon receiving Institutional Research Board (IRB) approval for the study (2008) was to administer a survey in which we asked students open-ended questions about their understanding of scientific communication and had them complete a checklist to rate the importance of various components of scientific research articles. Given that in both courses students would be completing research articles based on their laboratory experiments, we thought it particularly important to gain an initial idea of students' beliefs about scientific knowledge making, then study how those beliefs changed over an academic year.

Based on responses to the initial survey, we invited several students to participate in follow-up interviews, offering $100 in compensation (funded by an internal grant). Our choice of potential participants to invite was guided by a range of responses to the survey: We wanted participants who seemed to be in different stages of their understanding of what it meant to communicate as a scientist as well as those who had made interesting comments in their responses.

In this first phase, we conducted two interviews with each student – at the start and end of the semester – using a focused interview protocol (also known as semi-structured interviews; see Lindlof & Taylor 2010) so that we could follow a general guideline for interviewing all participants but allow interviews to focus on understanding participants' perspectives. Interviews were divided into two parts: Part 1 included a common set of questions, which provided comparable data across interviewees and instructional contexts. Part 2 included questions specific to students in each course, so that we could probe students' experiences in EMB/SciComm and BE.

We also collected all student writing over the semester, allowing us to conduct discourse-based interviews (Odell, Goswami, & Harrington 1983) in which students talked us through their processes for completing course writing tasks and offered reflections on the challenges and successes of their writing as well as the larger role that writing and speaking played for them. That student writing also allowed us to triangulate our interview findings in that we could offer evidence of students' writing development over the course of a semester. Though we had deep familiarity with the instructional contexts for these CI classes, we also conducted start and end-of-term interviews with the communication instructors and course faculty, analyzed pedagogical materials (syllabi, assignment sheets, handouts), and observed several classes in which communication instructors or course faculty would offer instruction on the CI component of the course. The field notes we made during these observations provided sources of data to understand the larger teaching and learning context and material to refer to in student and instructor interviews.

In 2010, we obtained IRB approval again and contacted the focal participants. All three participated in hour-long, semi-structured interviews shortly before graduation. We also asked these students to submit, before the interview, an example of a writing or speaking task completed during the year that they felt was representative of their current work. These interviews were focused on students' experiences with scientific communications in the two previous years, their reflections on the initial CI classes, the ways that they expected writing and speaking to play a role in their futures, and the specific practices and processes they used to create the examples that they had submitted.

Data analysis

Given our longitudinal data set, we wanted to know how students' conception of scientific knowledge making and communicating had changed over the two years between interviews. Because all three students had changed majors, we also were interested in how issues of identity emerged. Thus, our analysis of interview data was guided by these areas of inquiry: scientific knowledge making, communication, and expressions of identity.

We coded the three interviews for each participant using Beaufort's (2007) model of five knowledge domains in students' writing development: discourse community knowledge, rhetorical knowledge, content knowledge, writing process knowledge, and genre knowledge. We also coded for three prevalent dimensions of identity that emerged from the initial study: professional identity, student identity, and *writerly* identity, or the ways that students described themselves as writers or communicators (see Appendix for examples of codes). Thus, by using Beaufort's knowledge development scheme, we engaged in "etic" coding, as we used existing theories or categories to describe our interview data, as well as "emic" coding in seeing identity issues as emerging as important to participants based on interviews (Lett 1990).

We used the software program HyperResearch ⟨http://researchware.com⟩ to code interviews and then report on, collate, and count the frequencies of coded data. The resulting quantitative data gave us ideas as to how students' expression of the eight codes changed over time, as evidenced in the example for Nedra in Figure 4.1, as well as enabled a comparison across the three participants.

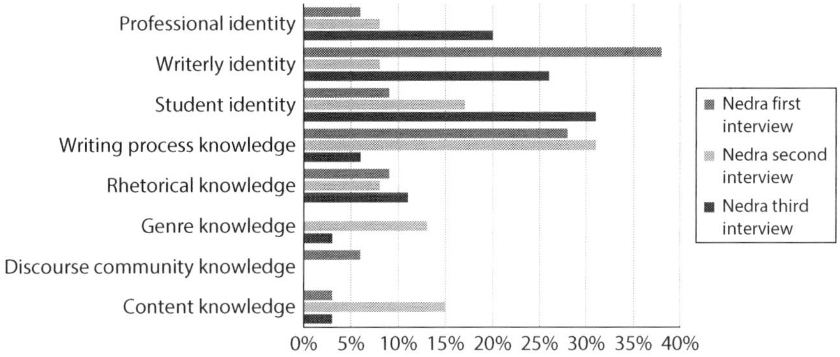

Figure 4.1. Example code frequencies in Nedra's interviews. Percentage refers to the ratio of the particular code occurrence to the occurrence of all codes (e.g. in Nedra's third interview, we applied the code "Professional Identity" to 20% of all codes that occurred)

The specific interview excerpts we coded (304 in all) offered nuanced and detailed examples of students' knowledge making and shifting identities; the results that we describe in the next section are based on these coded excerpts. In what follows, we offer profiles of each participant to explore the relationship between knowledge and identity making.

Case 1: April – Tentative steps toward the future

In 2008, April had been a chemical engineering major, taking EMB/SciComm as a required laboratory course; however, before her junior year and largely as a result of her SciComm experience, April switched to biology, the department from which she would graduate. Through the three sets of interviews over two years, April offered a relatively tentative vision of her future and by graduation she was still not sure of where she would be heading. The theme of not being completely in control of her life as a student marked April much more strongly than the other two participants. For example, April expressed surprise that she had been accepted to MIT, telling us that "My high school wasn't really that great. I mean, I wasn't even expecting to get into MIT at all, so, you know, it just kind of threw me." Like many students, she experienced a shift in what was interesting and what she was good at in high school versus what she encountered once she came to college. As she described,

> I sort of liked chemistry a lot in high school. Bio was OK. And I got here, and I did not like the chemistry classes, and I really liked the bio classes, but I wasn't sure if I wanted to be a pure scientist. So I said, OK, I'll do all of it. I'll do chemistry and biology and engineering. All of it, and it will be great, and it was awful.

April's move from chemical engineering to biology was largely guided, in her words, by what she did and did not find interesting and "fun" in what she was studying:

> [Chemical engineering] just wasn't interesting to me. It wasn't as much fun as I was hoping. It was more programming. It was more about what the machines can do rather than the science that's happening inside the machines. So I figured, of the two sciences that I'm not actually studying, I liked the biology classes more than the chemistry classes. So I figured that if those are the classes that I want to go to, I should make that my major.

In her final semester, April was enrolled in Project Lab, the second CI in the Biology Department, a course in which students are pursuing independent and novel research, akin to being a lab member of one of the two Biology Department faculty members who teach the course. In Project Lab, students give oral presentations every two weeks on their work research in progress, continually write up their research in the form of a scientific research article, and ultimately produce the equivalent of a publishable article by the end of the semester. It was this article that April chose as representative of her progress as a scientific writer as a senior.

In terms of developing knowledge about communicating science, as a sophomore April had been focused primarily on translating laboratory science in clear and concise ways. In her two interviews that year, she rarely mentioned a reader or audience for her writing, only the need to master particular formats or conventions. When she mentioned audience, it was with a great deal of uncertainty: "It's a good skill to have, in general, to be able to perform an experiment and get the results and be able to tell it to someone else. I guess that's definitely something good to know how to do."

Two years later, April's description of the task of communicating science was much more focused on recognizing the needs of the audience, that is, much more rhetorical. As she related: "I'm not saying that like I'm the best writer, but I think I know what someone who's reading my paper would be looking for, for example. And that's helpful, rather than just kind of going into it sort of blindly." This new-found rhetorical awareness was also reflected in the process April described for writing her Project Laboratory report, the document she chose for her third interview as reflective of where she was as a scientific communicator. An understanding of the audience beyond her classmates or professors gave her a starting point:

> It wasn't that difficult, I think, to make a broader audience than just the people in the class, just professors reading the paper, but … I definitely think it was a lot easier for this class to sort of focus in on the audience, who I think would be reading the papers.

Interestingly, April's new-found sense of rhetorical awareness was not primarily the result of experience with required communication-intensive courses. What she described as her most influential experience at MIT – one that she felt taught her to write – was taking a biology course in which no writing was required. Instead, students read scientific literature and presented on those readings in class. Critical reading – understanding what readers were and were not looking for and gaining an idea of the expectations of the discourse community in which she was a fledgling member – allowed April to write more effectively, anticipating readers' responses:

> It was an hour of lecture on Monday and then Tuesday was a two-hour class, and we would have two papers on whatever the topics was for that week, and people would present it and ask a lot of questions and sort of really look very critically at it. So it wasn't actually a writing class at all. But I think it helps a lot in thinking about what goes into a paper.

This key experience was one April came back to repeatedly in her last interview, including in response to the question, "Anything else to add?":

> I guess the most surprising thing for me was again just the [paper-reading course], just how much better I felt at writing. Like I know I feel a lot better at reading the papers. But I'm surprised by how much better I feel about writing them after that class, because there was no writing in there.

In a sense, this class offered April a critical lens through which to read scientific literature. As she described, "Going into that class at the beginning of this semester, just the kind of things I would look for in the paper were totally different from the kinds of things that they would ask us in class." She incorporated this critical reading lens into her writing repertoire, feeling better equipped to anticipate the questions and concerns of her scientific readers.

As a senior reflecting on her undergraduate experiences, particularly those having to do with scientific research and communication, April sounded far more in control of her learning than she had as a sophomore. By the time she was ready to graduate from MIT, April had experienced two semesters of intensive practice in scientific communication in SciComm and Project Lab, and another semester in her scientific reading course. All of this experience informed the following self-assessment:

> I'm pretty satisfied with where I am now, not to say I couldn't improve, but I think, I know I feel a lot more comfortable going up in front of the class and talking about something, whether or not it's something I did, or just something I read about in a paper and had to present.

Overall, April's move from chemical engineering to biology and her attempts to become a skilled scientific communicator were guided by a careful observation of what the scientists around her – particularly the biology faculty – were doing and the ways they critically read scientific texts. Her interest in particular areas of science as opposed to others guided her decision to change majors, and her growing sense of – and confidence in – herself as a scientific communicator stemmed from her opportunities to be engaged in authentic processes: conducting and presenting original research, reading and discussing scientific literature. Balancing the relationship between scientific content, communicative processes, and identity was an ongoing process for April, one that was by no means finished by the time of graduation.

For Carla, whom we next describe, such relationships also played key roles in determining her course of study and her post-graduation life though with far more surety and confidence than we saw with April.

Case 2: Carla – A teacher emerges

Between 2008 and 2010, Carla switched from biology to material science to political science as her major. One motivation for these transitions came during the summer before her junior year when she worked as a biology instructor for a program serving public school students. In the initial interviews, Carla had expressed a strong desire to teach biology and the summer experience solidified her commitment to teaching. She searched for a major that would allow her to pursue study of education and continue to build the analytical and research skills that had originally drawn her to MIT. As a political science major, Carla said, "I'm really happy in it. So it's like I can do research, I can do my research in education policy, and everything's good."

In terms of development as a scientific communicator, as a sophomore Carla had represented a fairly writer-based view of communicating science. As she noted in the initial survey, "I tend to write in a manner where the reader has to fill in the gaps." She also saw the task of scientific communication as largely concerned with clarity and conciseness, although her sense of audience was general: "You always have to state everything you did, state why you did it and then be clear and concise so anyone from any field can understand what you're saying." Unfortunately, this general approach was not aligned with the specific expectations of the SciComm research report, and as a result, Carla's grades on individual sections of the assignment were not high. By the end of the semester, she expressed frustration with her lack of experience: "SciComm was a little excruciating because I had never written like that before, so I was just like I don't really know what I'm doing."

As a graduating senior, Carla's understanding of the rhetorical situation for communicating science was much more nuanced and reader-based. The project she brought to the third interview as reflective of her writing development was a study on the voucher system in Milwaukee, WI, public schools, which she had written for a political science course. In terms of its literature review, she noted in her interview that "I write the lit review as though I'm explaining the entire concept to somebody, as though I'm standing in front of a bunch of people and then this is how this would happen."

However, rather than aim for a broad audience as she had mentioned two years earlier, by graduation, Carla's notions of readers' needs were built upon specific experiences presenting her work in political science classes, particularly in question-and-answer sessions after formal presentations:

> When I first came to [political science], and I gave my first presentation, people were asking me all these questions, and I was, so, like, I don't know, I didn't think of that, I didn't cover that or I didn't look that up. And now it's like nope, got to know everything, because you will get destroyed if you don't.

Carla had developed an understanding of communication that was beyond clarity and conciseness, but instead was about connecting to an audience, particularly when giving presentations:

> I feel like that's really important when you speak to just at least connect; you don't have to like be the most charismatic in the room, but you definitely have to just form a connection to the point where they are interested in your speaking, as well as just [in] the data.

For Carla, this need to form a connection was one she had experienced particularly as a teacher, the key influence on who she was now as a writer and speaker. When Carla described her major communication tasks as a senior, she noted, "We had one group presentation and one individual presentation this year. And I don't really have a problem with them especially after teaching, it's just kind of like giving a lesson." Along these lines, she also said, "The one thing that I think I learned more with teaching than at MIT is connecting with your audience."

Her experiences as a teacher also helped Carla become "more accommodating, more understanding and empathetic" toward others' learning. In her interview as a senior, she felt that she had not brought these qualities to MIT but had developed them through her struggles in biology and to find a major and an academic niche. In a sense, her student identity underwent a radical transformation:

> When I was in high school, I was totally that kid that was, like, why doesn't everybody get this? ... I hung out with all the other AP students and honor roll kids, and we're like, why don't, I just don't understand how no one else is understanding this, and, like, the stuff is easy. And here that was completely blown out, like that was done, like that part of life is over.

Carla also felt that the support, flexibility, and individual attention she received from political science faculty were key to her learning, whether technical content or communication skills. She contrasted that attention with her perception of biology, where she felt learning was less individual, particularly in terms of individual contact with her instructors.

Carla's path – from biology to material science to political science – did bring about a certain identity crisis. As she noted as a senior, few undergraduates come to MIT to major in political science, despite the department's stellar reputation for undergraduate research opportunities and teaching quality. When she contemplated becoming a political science major, she knew she was rubbing up against the perceptions of her peers: "A lot of people just like don't take [political science] seriously, and for me that was like one of my big troubles when I decided to switch, I was like, oh, am I actually going to be taken serious as like a researcher?" Once Carla was immersed in the work of political science and saw that analytical models and statistical analysis were central to the research she would be doing, she found comfort in this combination of "hard" and "soft" skills and peace with that student identity. Thus, at graduation she noted, "I do feel like a political scientist, especially after these classes that teach you all the rigor that goes into doing the analysis."

Carla's identities as a researcher, political scientist, and teacher offered a variety of future paths. By the time of graduation, her plan was to sign up with Teach for America for two years and then decide if she would pursue a career as a middle school teacher or get a Ph.D. in political science. Whatever the path, Carla felt that

MIT had prepared her well as a communicator and political scientist: "The best thing is like I feel prepared to walk into any job and be able to present or be able to write a paper on it. I feel extremely confident doing that."

In sum, Carla's movement through several majors at MIT was driven by her personal interests and her success as a communicator. In her final choice – political science – she could reconcile her identity as a researcher, her student identity, and her identity as a teacher, testament to the complex interactions between identity, content, and the discourses of a particular field. In the next case, Nedra similarly worked to balance multiple identities and justify her choice of final major.

Case 3: Nedra – Timing is everything

At the time of our initial study, Nedra was a sophomore biological engineering major and we tracked her experiences writing for the BE course. Nedra was also in one of the initial cohorts in biological engineering in 2006. This "newness" was one reason that Nedra decided to change majors after her BE experience. As she described, "biological engineering was a new major, especially for our class. For the freshman now, it's not as much of a new major. But a lot of the classes were very experimental and not necessarily established." Nedra particularly felt that BE faculty did not have a clear sense of students' schedules and were not as flexible as she would have liked in terms of assignment due dates and examination scheduling. Further, Nedra made a similar calculation as did Carla, looking ahead at the biological engineering curriculum and not seeing a good fit based on her interests: "I really liked my research, but I was looking at the classes that I had ahead of me. And I didn't seem very excited except for about two of them. So I realized that if I wasn't excited about what I was doing, then maybe I was in the wrong major." Two years later, Nedra was graduating as a mechanical engineering major, having made this switch at the start of her junior year.

In terms of developing knowledge of scientific communication, when Nedra described the act of communicating science as a sophomore, she indicated an awareness of audience and context for that communication and the idea that the purpose of such communication was collaborative, not competitive:

> I think it's important to get your message out there, but also it's important to kind of throw your ideas out and see the feedback. And I feel like that's a little harder to do because scientists tend to get at you, and not exactly attack you, but it depends on how you receive criticism. But it's a great way, speaking and writing is a great way to just collaborate ideas, see what's going on and get farther away?

A strong current throughout the initial interviews, however, was Nedra's sense that time was far too short in classes and did not reflect the authentic communication situations that the assignments were trying to address. As she described,

> I have conflicting feelings about [BE], because I love research. I absolutely love research. And I feel like in the real world you get a little longer of a timeline for coming up with an idea to refactor a genome or writing up a lab report or coming up with a research proposal. So I mean, I feel like I'd like real life better, but this was kind of real life or your entire research career compacted into one semester.

Thus, for Nedra, her sense of controlling a rhetorical situation was constrained by the lack of time as well as her feelings of having poor writing abilities as compared to classmates. For example, much of Nedra's writing in BE received relatively poor grades because she "ran out of time" when addressing instructor comments on initial drafts.

As a graduating senior, however, Nedra portrayed far more confidence in herself as a communicator, one who had mastered time or learned to apportion time more strategically than she had as a sophomore. By senior year, she had figured out the formula: "So now when I get a paper, I know how many, like, based on how many pages. I know how long it'll take me to write it, instantaneously."

As far as her choice of writing task as representative of the work she did as a senior mechanical engineering major, Nedra submitted her senior thesis, which was a study of the career trajectories and reflections on their undergraduate experiences of MIT alumni of the two undergraduate tracks of mechanical engineering – one primarily research-based with a fairly rigid curriculum (known as "Course 2") and the other offering more flexibility and choice of electives (known as "Course 2A"); Nedra had been in the latter track. While completing a thesis was optional, Nedra felt motivated to pursue this project, conducting extensive research on the discourse community that she had joined and, in a sense, justifying her choice and her identity as a student. As she described in the third interview,

> I also just understood that a lot of people really like 2A, but, and since I switched into 2A, a lot of Course 2 students would be, like, "oh, you're in the non-hardcore." And I get made fun of to this day, like, I'll, like, say something. And my friends will be, like, "right, but you're 2A." And it won't even have anything to do with what I'm saying. But I, like, lose credibility. And I mean, it's all in jest. It's not serious. But sometimes I do get a little pissed off, because I'm, like, all right, you know, this is kind of ridiculous.

As noted, Nedra's response to her peers' assertion of her identity in her major was to investigate graduates' experiences, and she found that post-graduate and career options were not limited by one's choice of mechanical engineering strand. Nedra's research also allowed her to justify the utility of learning the "soft skills" of collaboration and communication, which she felt her final course of study emphasized.

On the cusp of graduation, having secured a job in the field of energy consulting, Nedra could easily imagine the rhetorical situations and discourse community she would enter post graduation:

> There'll probably be a lot of data … behind a lot of the decisions we make. And while we can give them the decisions, they might want to go back into the data, especially afterwards, retrospectively to see, oh, you know, like, what, if we change this or X or Y, you know, what would change? And they'd have to look back at the data. And usually this data's never been summarized in the way that we would be doing it.

For Nedra, the "we" identity that could map on to these "real world" communication contexts came from the foundation she established as an undergraduate. As a graduating senior, she reported a high degree of confidence, feeling that the number of opportunities for teamwork and, in particular, presentations at MIT had prepared her well:

> I think that MIT's made it so that I'm not really afraid of anything. So if I were to begin a new task in a field that I've never dealt with before, or to present in front of an audience of 500 people, no matter what, it's a daunting task. It'll take a lot of time. And I'll probably be stressed. But I'm not afraid.

In sum, Nedra's experiences as a scientific communicator were shaped by her content interests, her perceived limitations as a writer, and her transition from a student identity to a professional identity. We believe the interplay of these factors for all students is central to the processes of learning to communicate in technical fields, particularly given the curricular emphasis on communication tasks that mirror professional tasks and the ways that students enter the larger discourse communities of scientific practice.

Conclusions, implications, and limitations

April, Carla, and Nedra's experiences attest to the complexities of learning to communicate science and to the interplay of identity and knowledge making. We want to draw attention to two particular conclusions we see emerging from these students' experiences, ones we believe are central to the contexts of schooling and that have important implications for teaching:

1. Entering the discourse community through its communicative practices: All three participants underwent a shift in student/writerly/professional identities between sophomore and senior years. These shifts had a great deal to

do with their perceived success (or lack thereof) with communication in their majors as well as with interest in and mastery of the field's technical content. Thus, by the time they were graduating, April, Carla, and Nedra could envision themselves as contributing members of discourse communities – not the communities that they entered but the communities that ultimately became their disciplinary homes. Their discursive experiences helped them reject those initial communities and shaped their membership in new communities.

Much of the scholarship on writing development in the disciplines looks at the uneven trajectory of students within majors, but our research suggests that it is equally valuable to investigate failed initiation into discourse communities – how and why students leave disciplines, in part, because they do not identify with the discursive practices of the discipline. Our research also suggests that we look to the surrounding structural difficulties that may compound newcomers' initiation into disciplinary communities. Lack of interpersonal support or stifling curricula requirements may undermine authentic writing instruction. Such structural and even macro-structural issues may be especially important for women in the sciences. In such cases, students' frustration with writing is like the canary in the coalmine, signaling deeper problems in major. Likewise, writing researchers too commonly ignore the significance of reading in the disciplinary initiation process; it is important to note that for April, membership in her academic discourse community came more from reading research literature and anticipating what readers of future research reports might expect, rather than from a writing course per se. In a sense, her "legitimate peripheral participation" (Lave & Wenger 1991) gave her a view of expert practices and built the comfort she needed to make those practices her own.

2. Managing the limits of authenticity in school contexts: As sophomores, all three students had fairly unsophisticated beliefs about communicating science and they could not imagine the rhetorical situations and discourse communities for what and whom they were writing beyond their instructors. As mentioned, this constraint on *authenticity* of the classroom setting has been noted by several researchers (e.g. Artemeva 2009; Dannels 2000; Freedman & Adam 2000). Nedra, in particular, chafed at the school-based constraints she experienced.

   By the time they were graduating, April, Carla, and Nedra had become "active observers" and participants in knowledge production, all feeling confident about their preparation for the communicative challenges ahead. As they map their future identities onto the kinds of writing and speaking they believe they will do, they tell us a great deal about the experiences that precede those maps and offer lessons for future undergraduates engaged in the creation and communication of scientific knowledge.

The implications we want to draw attention to are what these students' experiences might tell us about teaching and curriculum. An important question for teaching disciplinary communication is how to engage students in these kinds of opportunities and help them develop the discursive identities that support success. Certainly, one fruitful way is to continue to embed communication instruction within courses in which students are learning technical content, as MIT and many other institutions attempt to do with CI courses. However, students may also have such learning opportunities outside of recognized or sanctioned "communication-intensive" courses. Thus, to make students' development of communicative expertise a goal in undergraduate science requires a broad mapping of opportunities for communication activities across a wide variety of courses throughout the curriculum rather than restricting opportunities to designated CI courses or capstone experiences, which typically involve writing and presenting.

Another question for teaching is how to develop authentic reading and writing tasks when students have little sense of what authentic communication practices might entail or audiences for those practices, as was generally true for our participants as sophomores (Curry 2012; Ding 2008). One answer comes from what is known about the intertwined relationship between knowledge creation and knowledge reporting and the necessity to have students participate in both activities. As Connolly (1989) describes,

> School serves two valuable and quite different functions. It exists to share conventional knowledge-to indoctrinate new members into the operative beliefs of a community: its scientific, social scientific, humanistic, and artistic body of learning. But it exists also to initiate passive participants in the common life into active observers who understand what knowledge is and how it is made, and can participate in its production. (p. 2)

The goal of students as "active observers" and participants in knowledge production is perhaps assumed to occur in curricula in which students are engaged in hands-on laboratory learning. However, just as some school-based laboratory practices are more authentic than others (Hodson 1998), some communicative tasks offer students more authentic experiences than others True opportunities to produce and communicate new knowledge are challenging to curriculum planners and teachers themselves (for one example, see Hanauer, Jacobs-Sera, Pedulla, Cresawn, Hendrix, & Hatfull 2006), and as we've pointed out, the contexts of schooling are a significant limitation to "authenticity." Nevertheless, we see class and laboratories as an opportunity to focus student attention and put into place novice-expert models of learning that might not be available in professional or workplace contexts, particularly in terms of explicit instruction. Such explicitness is key here, whether that's an exploration of the rhetorical situation, the motivations of instruction, or the role of writing, speaking, and visualizing to create and communicate knowledge.

Ultimately, we see a continuing need to better understand student learning through *students'* perspectives. In the current era of outcomes assessment and the push to quantify student learning, we need to remind ourselves continually of the importance of students' stories of learning and becoming. These stories, particularly when they occur over a significant span of an undergraduate's career, as in this study, can tell us a great deal about the results of teaching and curricular designs and, ultimately, guide efforts at reform, especially the development of initiatives such as communication-intensive curricula.

Of course, our conclusions, based on the research we report here, are tempered by several factors, many inherent to qualitative research:

- We present the experiences of only three undergraduates and make no claims that these three are representative of all MIT undergraduates, much less undergraduates at other institutions. All three of our participants were women of color; however, we did not use a theoretical frame that accounted for that racial or gendered identity as our larger study did not include such a frame. This oversight meant that we lost a potentially rich frame of analysis.
- Our methodology primarily relies on our understandings of the courses and tasks that students were engaged in and students' accounts of their experiences with learning to communicate science, rather than direct observation of those experiences, creating the possibility for inaccuracy or miscommunication. As Mishler (1990) describes, "the account produced during the interview is a reconstruction of the past, shaped by the particular context of its telling" (p. 427).

In the end while MIT's context affords resources unavailable at many other institutions, the issues our participants faced are not unique to MIT. Our conversations with writing-across-the-curriculum colleagues – in the sciences and engineering and more broadly – have shown how common are the challenges for students to develop disciplinary expertise as communicators and for institutions to create instruction that allows students to succeed in this goal. Thus we believe our study holds implications for considering these challenges for a wide variety of students across a range of institutions and it is incumbent upon us as educators and researchers to recognize and better understand their efforts and develop teaching methods that support such development.

References

Artemeva, N. (2005). A time to speak, a time to act: A rhetorical genre analysis of a novice engineer's calculated risk taking. *Journal of Business and Technical Communication, 19*(4), 389–421.

Artemeva, N. (2009). Stories of becoming: A study of novice engineers learning genres of their profession. In C. Bazerman, A. Bonini, & D. Figueiredo (Eds.), *Genre in a changing world: Perspectives on writing* (pp. 158–178). Fort Collins, CO: WAC Clearinghouse and Parlor Press.

Beaufort, A. (2007). *College writing and beyond: A new framework for university writing instruction.* Logan, UT: Utah State University Press.

Blakeslee, A. (2001). *Interacting with audiences: Social influences on the production of scientific writing.* Mahwah, NJ: Lawrence Erlbaum Associates.

Brown, B.A., Reveles, J.M., & Kelly, G.J. (2005). Scientific literacy and discursive identity: A theoretical framework for understanding science learning. *Science Education, 89,* 779–802.

Carroll, L.A. (2002). *Rehearsing new roles: How college students develop as writers.* Carbondale, IL: Southern Illinois University Press.

Connolly, P. (1989). Writing and the ecology of learning. In P. Connolly & T. Vilardi (Eds.), *Writing to learn mathematics and science* (pp. 1–12). New York, NY: Teachers College Press.

Curry, M.J. (2012). Transcending 'traditional academic boundaries': Designing and implementing a science communication course for science and engineering Ph.D. students. *Professional and Academic Writing, 40,* 4–7.

Dannels, D.P. (2000). Learning to be professional: Technical classroom discourse, practice, and professional identity construction. *Journal of Business and Technical Communication, 14*(1), 5–32.

Ding, H. (2008). The use of cognitive and social apprenticeship to teach a disciplinary genre initiation of graduate students into NIH grant writing. *Written Communication, 25*(1), 3–52.

Ellis, R.A. (2004). University student approaches to learning science through writing. *International Journal of Science Education, 26*(15), 1835–1853.

Freedman, A., & Adam, C. (2000). Write where you are: Situating learning to write in university and workplace settings. In P. Dias & A. Paré (Eds.), *Transitions: Writing in academic and workplace settings* (pp. 31–60). Cresskill, NJ: Hampton Press.

Haas, C. (1994). Learning to read biology: One student's rhetorical development in college. *Written Communication, 11,* 43–84.

Hanauer, D.I., Jacobs-Sera, D., Pedulla, M.L., Cresawn, S.G., Hendrix, R.W., & Hatfull G. F. (2006). Inquiry learning: Teaching scientific inquiry. *Science, 314*(5807), 1880–1.

Haswell, R.H. (1991). *Gaining grounding ground in college writing: Tales of development and interpretation.* Dallas, TX: Southern Methodist University Press.

Herrington, A. & Curtis, M. (2000). *Persons in process: Four stories of writing and personal development in college.* Urbana, IL: National Council of Teachers of English.

Hodson, D. (1998). Is this really what scientists do? Seeking a more authentic science and beyond the school laboratory. In J. Wellington (Ed.), *Practical work in school science: Which way now?* (pp. 93–108). London: Routledge.

Lave, J., & Wenger, E. (1991). *Situated learning: Legitimate peripheral participation.* Cambridge: Cambridge University Press.

Lett, J. (1990). Emics and etics: Notes on the epistemology of anthropology. In T.N. Headland, K.L. Pike, & M. Harris (Eds.), *Emics and etics: The insider/outsider debate.* Frontiers of Anthropology, 7 (pp. 127–142). Newbury Park, CA: Sage.

Leydens, J.A. (2008). Novice and insider perspectives on academic workplace writing: Toward a continuum of rhetorical awareness. *IEEE Transactions on Professional Communication, 51*(3), 242–263.

Lindlof, T.R., & Taylor, B.C. (2011). *Qualitative communication research methods* (3rd ed.). Los Angeles, CA: Sage.

Mishler, E.G. (1990). Validation in inquiry-guided research: The role of exemplars in narrative studies. *Harvard Educational Review, 60*, 415–442.

Odell, L., Goswami, D., & Herrington, A. (1983). The discourse-based interview: A procedure for exploring the tacit knowledge of writers in nonacademic settings. In P. Mosenthal, L. Tamor & S. Walmsley (Eds.), *Writing research: Methods and procedures* (pp. 221–236). London: Longman.

Patton, M. (2011). *Writing in the research university: A Darwinian study of WID with cases from civil engineering.* Cresskill, NJ: Hampton Press.

Poe, M., Lerner, N., & Craig, J. (2010). *Learning to communicate in science and engineering: Case studies from MIT.* Cambridge, MA: The MIT Press.

Russell, D. (2001). Where do the naturalistic studies of WAC/WID point? A research review. In S. McLeod, E. Miraglia, M. Soven, & C. Thaiss (Eds.), *WAC for the new millennium: Strategies for continuing writing across the curriculum programs* (pp. 259–298). Urbana, IL: National Council of Teachers of English.

Sommers, N., & Saltz, L. (2004). The novice as expert. *College Composition and Communication, 56*(1), 124–149.

Thaiss, C., & Porter, T. (2010). The state of WAC/WID in 2010: Methods and results of the U.S. Survey of the International WAC/WID Mapping Project. *College Composition and Communication, 61*(3), 534–570.

Yin, R.K. (2008). *Case study research: Design and methods* (4th ed.). Thousand Oaks, CA: Sage.

Appendix

Examples of Coded Interview Excerpts

Code	Example
Content knowledge	April: "Going into this paper, I know the best I can about writing it, but I know from doing the experiments what we could have done better."
Discourse community knowledge	Carla: "But the same reason a poli sci major might think something is important is probably not the same reason that the math public might [think] so."
Genre knowledge	Nedra: "One of the comments was I didn't have enough detail. I mean, I didn't really know how much detail to put in. I wanted to be concise. But I guess I was too concise."
Rhetorical knowledge	April: "It's a good skill to have, in general, to be able to perform an experience and get the results and be able to tell it to someone else, I guess that's definitely something good to know how to do."

(*Continued*)

Appendix. (Continued)

Code	Example
Writing process knowledge	Carla: "I would say it was like compared to a regular like humanities-type paper. I outlined more to the point where pretty much the outline was the paper, and I would just change the way I stated my bullet points, I suppose. Because it was just very methodical, everything was, like, and then we did this and then we did this, so it was extremely, extremely structured."
Student identity	April: "My high school wasn't really that great. I mean, I wasn't even expecting to get into MIT at all, so, you know, it just kind of threw me."
Writerly identity	Carla: "I feel comfortable with the type of language I'm supposed to use, exactly what I'm supposed to include in my paper to make it understandable to other people. Pretty much everything about writing a journal article or a lab report, which I guess it was kind of just a lab report I feel very comfortable with."
Professional identity	Nedra: "With consulting, you do an incredible amount of number crunching and then summarizing that data in the form of reports. So I will be writing a lot. I will be presenting a lot. And I will definitely be involved in also reading technical information, being able to assimilate that, especially when I get involved in policy."

PART II

Research studies in engineering

CHAPTER 5

Writing as social practice in engineering

Views from a cross-disciplinary study

Julio Gimenez
The University of Nottingham

This chapter reports on a collaborative study that used a range of qualitative data (interviews, observations, talk around text, and written reports) as its empirical base to examine academic and professional writing in four schools of the faculty of engineering at a British university. Findings demonstrate that writing in these schools is closely related to its context of production. As such, writing not only reflects the views on knowledge and ideologies of engineering as a discipline but also helps to reaffirm them. The findings contribute to understandings of the nature and dynamics of writing as social practice that emerged from the research efforts of a cross-disciplinary team of researchers in applied linguistics and engineering. They show that both applied linguists and engineers can inform pedagogical approaches to disciplinary writing practices so that they can become more context- and discipline-specific. The chapter concludes by suggesting avenues for further cross-disciplinary research in writing.

Introduction

Writing in the disciplines has been investigated for quite some time, resulting in a large body of knowledge emerging from studies that have taken a disciplinary or writing/composition perspective (e.g. Alley 2000; Kim & Severison Eklundh 2001; Luzon 2005; Parks 2001; Seifert 2009; Winsor 2001; Zhu 2004). In engineering, writing has been recognised by engineers and writing experts to be a central activity in both the academy and industry, occupying a considerable amount of time in the professional activities of engineers (Bracewell & Witte 2003; Gygi & Zachry 2010; Nelson 2000).

However, few studies have examined academic and professional writing in engineering as social practice in which disciplinary values and ideologies are both created and reaffirmed (but see, e.g. Gimenez & Thondhlana 2012; Luzon 2005). This approach to disciplinary writing is essential for developing a nuanced understanding of the role and dynamics of writing in the specific contexts where it occurs. As Bazerman and Prior (2003, p. 2) argue, "to understand writing, we need to explore the practices that people engage in to produce texts as well as the ways that writing practices gain their meanings and functions as dynamic elements of specific cultural settings." Bazerman and Prior's argument thus identifies the need to look at disciplinary writing from both an applied linguistics perspective through which we can analyse the production of texts and a disciplinary perspective through which we can examine the meanings and purposes of writing as situated practice.

Cross-disciplinary research between applied linguists and engineering educators can provide an excellent opportunity for unpacking the nature and dynamics of disciplinary writing in the way suggested by Bazerman and Prior. It can also help to inform our understanding of writing practices in engineering and develop pedagogical approaches to disciplinary writing that can be beneficial to both groups of educators as well as students.

In the following sections, this chapter presents a brief discussion of writing as social practice, followed by an examination of the relationship between disciplinary epistemology, ideologies, and writing in engineering. Next, the chapter describes the aims, context, and methods of the study reported on here and discusses the issues and challenges of cross-disciplinary research. It then presents its main findings and discusses them in relation to how they can inform collaborative ways of researching and teaching disciplinary writing in engineering. The chapter ends with suggestions for avenues for further research in disciplinary writing.

Writing as social practice

Professional and academic writing has been investigated in a number of disciplines: business and economics (Seifert 2009; Zhu 2004), engineering (Bracewell & Witte 2003; Gimenez & Thondhlana 2012; Gygi & Zachry 2010; Kim & Severison Eklundh 2001; Luzon 2005; Winsor 2001), health sciences (Gimenez 2008; Parks 2001), and the natural sciences (Alley 2000), highlighting the central role it plays in academic and professional worlds. A number of theoretical and analytical perspectives have been adopted. Some studies have approached writing as a technical skill (Brieger 2011; Storch & Tapper 2009; van Emden 2005), others as a textual practice (Alley 2000; Seifert 2009; Wheeler & McDonald 2000; Zhu 2004), whilst others have looked at aspects beyond the textual and examined the pragmatics and politics of writing within a discipline (Gygi & Zachry 2010; Luzon 2005).

One useful way of capturing the textual, the pragmatic, and the political in writing is by conceptualising it as social practice, that is, as a purposeful social activity in which people engage to achieve a variety of aims in their interactions with others (Lillis 2008; Street 2001). In this way, writing can be examined in the specific contexts where it occurs, involving social and institutional constraints, and issues of individual as well as professional identities (Ivanic 1998). At the same time, examining writing as social practice allows us to develop a better understanding of the relationship between text and context (Lillis 2008). Instead of looking at writing as a technical skill, a social practice approach sees it as "embedded in socially constructed epistemological principles" (Street 2001, p. 2), as will be discussed in the following section. Thus, knowledge, social meanings, and texts constitute a social space in which a group of people, the contexts where they interact, the beliefs and values they uphold, the acts they engage in, the tools and technologies they use, and the discourses and words they recognize are all dynamically networked with one another (Barton, Hamilton, & Ivanic 2000). Theorising writing in such a way provides an opportunity to examine both the practices involved in the production of context-specific texts as well as the values and ideologies that shape the meanings of such practices.

Disciplinary epistemology, ideologies and writing in engineering

In the context of higher education, researchers following a social practice approach to writing have recently called attention to the role played by disciplinary epistemologies in shaping professional values and ideologies, and how these values and ideologies in turn shape professional practices such as writing (e.g. Gimenez 2012; Luzon 2005). Jones (2009), for example, has urged us to consider epistemology as a central element in disciplinary culture as the particular values, ideologies, and ways of thinking of a discipline are largely influenced by its main theories or beliefs about knowledge. In engineering, the values, ideologies and ways of thinking shape disciplinary practices such as writing design reports and presenting research findings in ways that are specific to engineering, giving rise to what some researchers refer to as "writing and speaking like an engineer" (Tenopir & King 2004; Winsor 1996) despite sub-disciplinary variation identified by other researchers (e.g. Curry 2012). Through such practices the members of the discipline are seen not only to create but also to help reaffirm the values and ideologies of their professional community.

Disciplinary epistemology is, however, a highly complex concept which should be understood as fluid, dynamic, and diverse, reflecting the nature of contemporary disciplines (Gimenez 2012; Jones 2009). Engineering serves as a clear example of the multifaceted nature of disciplinary epistemology – a combination of abstract and concrete knowledge that has shaped the central ideologies of the discipline – mainly reductionism and holism (Sherwin 2010; Wang 2008). Throughout its history, engineering has benefitted from the contributions of basic sciences, as in work by Henry (1886), engineering sciences (Isherwood 1863), and design which, according to McClellan (1913), was the responsibility of the "real engineer" (p. 1272). These fields of knowledge have contributed to the epistemology of engineering that combines abstract knowledge concerned with theoretical physics from the basic sciences with concrete knowledge focused on real-world problems from the design sciences. This way of defining knowledge particular to engineering has influenced the main ideologies in the discipline which also present a combination of reductionism and holism, as illustrated in Figure 5.1. Whilst reductionism advocates the principle of breaking down systems into constituent parts in order to be reconstituted into a single whole solution at a later stage, holism attempts to capture the essence of systemic behaviour rather than that of the individual parts of a system (Sherwin 2010; Wang 2008). The multifaceted nature of epistemology in engineering with its particular ideology has at the same time shaped the nature and dynamics of disciplinary writing, making it reflect principles and theories from both basic and design sciences in order to solve real-world problems, as I discuss below.

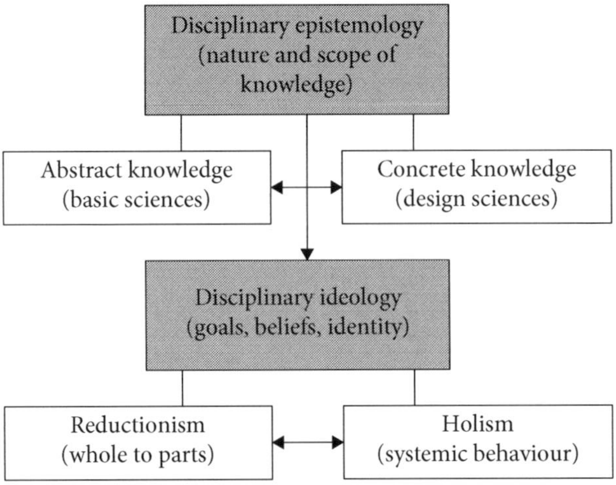

Figure 5.1. The nature of disciplinary epistemology and ideology in engineering

However, disciplinary epistemology is never an either-or concept. Recently, Figueiredo (2008) has shown that contemporary knowledge in engineering has resulted from the combination of four dimensions that include: basic sciences that have contributed particular analytical and positivist views of knowledge; human sciences that have brought hermeneutical, interpretive, and integrative traditions; design that has contributed functional analysis, problem-setting and problem-solving, and evolutionary learning within the framework of constructivism; and crafts that have shaped knowledge in engineering with pragmatist and reflective philosophies. Taking these four dimensions into account, we can see how the aggregative nature of engineering epistemology has helped to shape the values and ideologies of the discipline by defining it as "the creative application of scientific principles" (Engineers' Council for Professional Development, 1947, p. 456).

The fluid nature of knowledge and ideologies in engineering has also been reflected in the social practices of the discipline. Writing design reports, for instance, requires the combination of, *inter alia*, analytical and positivist contributions through the calculation and flow sheets needed for making design decisions and a holistic and integrated perspective on reality in order to solve real-world engineering problems. Thus, a nuanced understanding of writing design reports can be best achieved by examining it as a social practice rather than a textual exercise and within the context of cross-disciplinary research.

Engineering has a long history of cross-disciplinary and collaborative research. Projects between engineering, biological sciences, chemistry, education, health care, mathematics, and physics, to name a few, have been supported and, to a certain extent, encouraged by science foundations (e.g. the National Science Foundation in the United States), research councils (e.g. the Engineering and Physical Sciences Research Council in the United Kingdom), and universities (see, e.g. Christie, Ferdos, & Adawi 2009). However, collaborations between engineers and applied linguists have been less common (but see Gimenez & Thondhlana 2012). In cross-disciplinary examinations of writing, applied linguists can provide conceptual and analytical tools for looking at how writing as social practice embodies knowledge, values, and ideologies. At the same time, engineering educators can offer theoretical and analytical frameworks for understanding the meanings and functions of writing practices in specific social settings.

Cross-disciplinary examinations can be both enriching and challenging, especially between researchers from quite different disciplines such as engineering and applied linguistics. In the following sections, I present the participants and methodology of the study reported on this chapter, and discuss some of the main benefits and challenges of those collaborations in relation to the study.

Research methodology

Aims, context, and participants in the study

The study aimed to examine the nature and dynamics of writing as a social practice at four schools (Chemical and Environmental Engineering, Civil Engineering, Electrical and Electronic Engineering, and Mechanical, Materials and Manufacturing Engineering) of the faculty of engineering of a British university. Specifically, the study sought to develop a nuanced understanding of the practices involved in the production of specific texts across the faculty and the ideologies that shaped the meanings created by such practices. The goal of establishing this refined understanding was for it to be available to inform the practices of researching and teaching disciplinary writing for both applied linguists and engineering educators. The study was carried out by a team of four researchers: two applied linguists from the School of Education who have a background in academic writing and two engineers from the Faculty of Engineering who specialise in chemical and environmental engineering. Members of the team share an interest in cross-disciplinary collaborative research in general and the pedagogy of disciplinary writing in particular.

The study followed four phases. Phase I involved planning and designing the project, exploring and negotiating team members' views on disciplinary writing, and selecting the methodology, including data collection methods. Phase II involved recruiting undergraduate students and academics by invitations sent through various distribution lists facilitated by the engineering faculty. Phase III consisted of collecting a range of data (see description, below), and organising data sets for analysis. Preliminary data analyses were carried out by the applied linguists followed by discussions of emerging results by the whole team (see methodology, below). Phase IV involved making final decisions about the main findings, selecting data as supporting evidence, and dividing responsibilities for writing reports from the project.

Twenty-one undergraduate engineering students and four engineering academics, two of whom had substantial experience in industry, as described in Table 1, agreed to participate in the study. Before data were collected, the aims of the study were explained to the students and academics, their rights and issues of anonymity and confidentiality were discussed, and their consent to participate sought.

Table 5.1. Participating engineering students and academics

School in the faculty	Students	Academics	
		University	University/Industry
Chemical Engineering	6	1	
Civil Engineering	4		1
Electrical and Electronic Engineering	4	1	
Mechanical, Materials and Manufacturing Engineering	7		1
Total	21	2	2

Data collection and collaboration

Given its aims, the study was framed within a qualitative paradigm. The study aimed to characterise participants' experiences of writing as practised across the schools within the faculty and to capture how they made sense of the complexities involved in disciplinary writing as shaped by disciplinary epistemologies and ideologies. To analyse the data sets, we followed the analytical principles of critical grounded theory (CGT), a version of grounded theory proposed by Charmaz (2005, 2011) as they are two different publications. CGT offers a useful set of principles by which researchers can perform "successive levels of data analysis and conceptual development" (Charmaz 2005, p. 507) in order to develop inductive theories. CGT focuses closely on the phenomenon under study and adds a critical and reflexive dimension to grounded theory, by which researchers can examine in detail the context investigated, the data collected and their understandings of the data, whilst acknowledging they are part of the research reality. As Charmaz (2005, p. 529) explains, "a grounded theory informed by critical inquiry demands going deeper into the phenomenon itself and its situated location in the world."

In CGT, data collection and data analysis inform each other: Analysis helps to focus subsequent data collection, and data collection helps to refine the analytical theories, a process referred to as theoretical sampling (Charmaz 2005, 2011) as they are two different publications. As researchers develop this relational, cyclical process between data analysis and collection, they develop abstract and refined ideas about participants, their actions, and meanings within a specific social space. The process of theoretical sampling, at the same time, allows researchers to check their analytical hypotheses against the data sets, paying attention to evidence that both supports as well as contradicts their hypotheses. In Charmaz's (2005, p. 508) words, "researchers portray their understandings of research participants' actions and meaning, offer abstract interpretations of empirical relationships, and create conditional statements about implications of their analyses." A representation of these methodological cycles is provided in Figure 5.2.

The data sources for the study consisted of two exploratory meetings with participants and a number of interviews, observations, analyses of texts, and talk-around-text sessions. The purpose of the two exploratory meetings – one with a group of final-year undergraduates and one with the two academics with experience in industry, was twofold. The research team aimed at inductively exploring general trends in the writing practices across the faculty (i.e. whether writing was mainly an individual or collaborative practice, the role of technology in producing written documents, the relationship between academics and students) and at deductively confirming or changing the theoretical views of the research team.

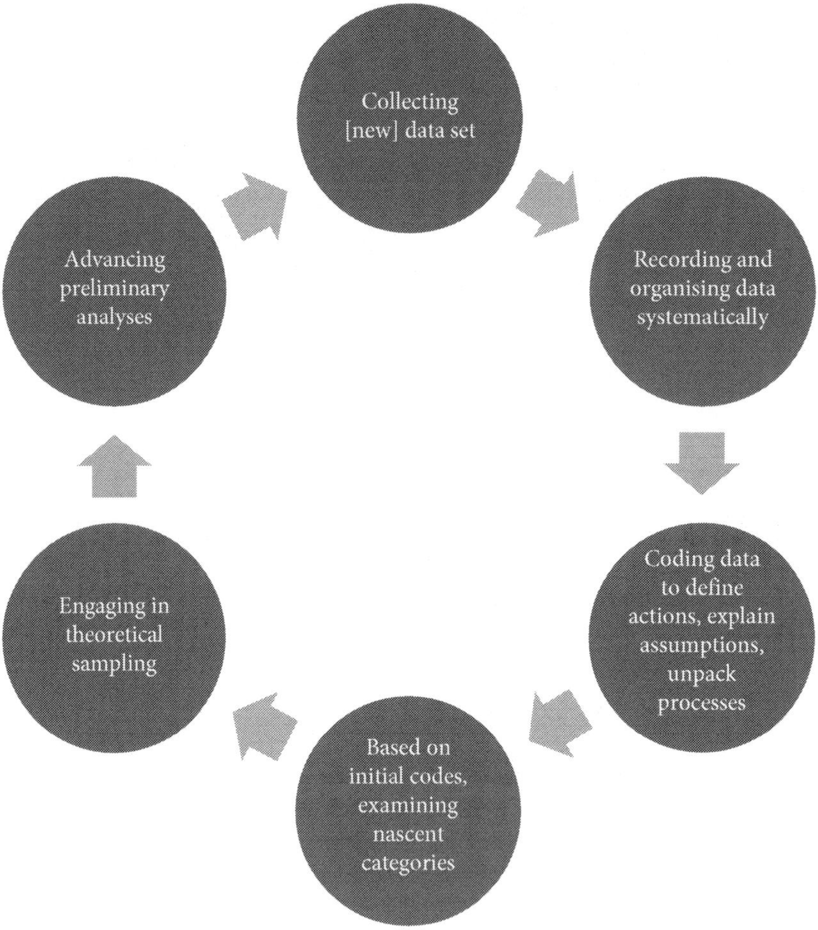

Figure 5.2. Methodological cycles

After analysing transcripts[1] of the meetings for emerging codes and nascent categories, and advancing our first set of analytical hypotheses, we conducted in-depth interviews with the four academics and the 21 participating students. The interviews were then transcribed. This time the applied linguists independently coded the transcripts, following the principles of what Strauss (1987) calls "open coding," with an aim to "produce concepts that seem to fit the data" (p. 27). The resulting codes, themes, and categories were discussed by the team and used to

1. Spoken data were recorded and transcribed verbatim. Transcripts were not edited for grammatical omissions or errors.

develop analytical interpretations of the main practices associated with writing in engineering. Next came observation and recording of the tutorials[2] where tutors and student groups discussed writing design reports. These recordings were transcribed and coded, with the resulting codes and categories used to refine existing analytical interpretations of writing as social practice. Once the students had submitted copies of their final design reports, we analysed those that had been discussed at the tutorial sessions. The results of the textual analyses of the reports were discussed by the team and used to refine our interpretations, enabling us to agree on preliminary findings.

These preliminary findings were examined during the 'talk-around-text' (Ivanic 1998; Lillis 2008) sessions with the student writers. Talk-around-text discussions take place between researchers and writers about the writers' texts in order to allow writers to explain key issues involved in writing a specific text that go beyond the textual. As Lillis (2008, p. 359) explains, talk around text "encourages comment and reflections that go beyond writing within current dominant conventions and practices and recognizes that the participants' analytic lens and perspectives are central to establishing what may be significant and important in any specific context."

Although the methodology for the study was decided on by the research team, the coding of most of the data was performed by the applied linguists. The resulting codes and taxonomy were analysed, discussed, and agreed on by the members of the research team before we embarked on the next round of data collection. The collection of data in stages allowed us to refine the analytical theories we developed in the research process. The discussion and decision processes were both enriching and challenging. Although we found the process of collaborating an exciting opportunity to gain insight into how the other members of the team constructed their understanding of writing, it was sometimes difficult to fully interpret the processes involved in such construction. Whilst the applied linguists tended to focus on the discursive construction of writing, the engineers emphasised the need for writing to reflect disciplinary knowledge. As discussed above, disciplinary views of the world and disciplinary positioning of academics are largely shaped by the epistemologies of their disciplines. Thus, as a team we had to engage in various rounds of discussion and examination of the findings to interpret them in a way that would fit our understandings of the phenomenon under investigation, from the positionings that we had developed as researchers in

2. In British universities, a tutorial is an individual or small group class in which a tutor (instructor) gives individual attention to students, usually in relation to a specific topic or task.

two very distinct disciplines, whilst trying to reach a collaborative understanding that would inform the study.

Another challenge was posed by our decision to distribute analytical tasks according to skill. Because the applied linguists were more experienced in data coding, it was decided that they would undertake the coding process and the results would be analysed, discussed, and agreed upon by all team members. As a result, collaboration depended mainly on making sure that we reached an agreement on the relative significance of the findings emerging from the analysis of the data sets.

Thus, collaboration required negotiating our interpretations of the emerging findings in an attempt to (1) reach a shared understanding of writing as situated practice (common ground); (2) develop common views on writing as a discursive process that embodies the central meanings (values) and specific functions (ideologies) in the discipline (collaborative views), which formed the basis for our analytical hypotheses; and (3) develop new ways of investigating and teaching disciplinary writing (new research and pedagogical perspectives). The interrelationship among these dimensions of the process is illustrated in Figure 5.3.

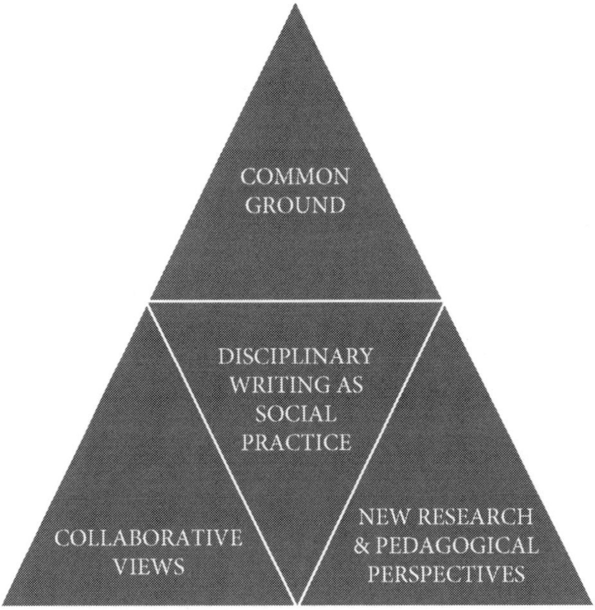

Figure 5.3. Dimensions in cross-disciplinary understanding of disciplinary writing

This engagement, however, made all of us more open to disciplinary interpretations and more aware of the relationship between disciplinary epistemologies and writing as a social practice where disciplinary values and ideologies are constructed, challenged, and reaffirmed (Gimenez 2012). Reaching a shared understanding of the nature and dynamics of writing in engineering was a necessary first step for the team members to be able to negotiate views on writing as a discursive process within the discipline, that is, as discursively constructed to reflect the values and ideologies of the discipline in order to become a vehicle for the creation and dissemination of disciplinary knowledge. This notion is explored in the next section.

Toward a nuanced understanding of writing as social practice in engineering

A significant finding that emerged from examining writing as social practice was the close relationship between writing and the various layers of context where it occurs: the faculty of engineering, the university, and engineering as a discipline. At the faculty level, writing was seen as a collaborative endeavour by a team of people who bring a range of skills and competences to the writing task. This view of writing as a collaborative effort is encapsulated in the views of the engineering academics provided in interviews:

> Well, you know, in a design project, for example, someone would typically be in charge of the calculations and the flow sheet, somebody else would be in charge of tests and reporting results, another member of the team would be writing the first draft and … someone who's a nit-picker would probably be the editor. (Int_02_Lect_CE)

> The work is group, I think that's the key, I mean there isn't an option really, nothing is, Brunel did it all himself maybe but nowadays nothing is done like that and so because you're working, all your work is as a group then inevitably the writing is going to be in a group. (Int_03_Lec_3Es)

> You are in a group, you are normally there because each person has something different to contribute. You know someone may be good at writing drafts but somebody else would probably do the editing, someone with a good eye for detail. (Int_04_Lect_3Ms)

As these extracts indicate, rather than being seen as a skill, here writing is conceptualised by the engineering academics as a purposeful social activity that requires interacting with others in order to achieve a common aim. This stance may explain the centrality of collaborative writing at the engineering faculty in this university, which may result from the particular ways the members of the engineering faculty construct academic and professional writing as well as from disciplinary ways of understanding writing for knowledge construction, as the following extract shows:

> In this particular school and I'd say at [name of university] we tend to work more collaboratively than in other places I know or worked before. I at the moment am writing a, you know, a kick-start to it [a project] and then that goes out around colleagues who come back and comment, we meet to decide and we decide where it's going next and somebody else might pick up and do the next bit as a starting point or it might be me again. And I maybe tend to then just refine ideas in that way, that's how we construct and shape our ideas too. We find it better to have something to argue against when we're writing it rather than this idea, everybody contributing something and trying to pick. (Int_01_Lect_ChE)

This way of understanding writing is also reflected in the type of writing practices in which engineering academics and students frequently engage: writing (design) projects, joint research articles, and collaborative research reports (Bracewell & Witte 2003; Gimenez & Thondhlana 2012; Gygi & Zachry 2010). Students who participated in the study seem to share this collaborative understanding of writing as seen in the following extract from a talk-around-text session on how a team of student writers had gone about writing a design report:

> Basically there's a list on there with everyone's section and what everyone did. As you can see … there's the names of the people who were actually involved in the actual design, the work involved, like acquiring the data for it. Find the writing on it, might make more sense to you. We were all involved in the completion but it was actually when it came to writing it up to bring it into one document it was finally written up by [name of team member]. See? What he did was, he took everyone's work and brought it together to generate this document basically. Then he sent it round to everybody. The list was reviewed by everyone towards the end of a project … we basically uploaded it onto an online data storage site where we had all had access to it and could post a modification if we wanted. (TAT_02_3Ms)

The student's explanation provides an interesting description of writing as a collaborative social activity. It refers to both the processes involved in achieving a common purpose (i.e. writing a design report) and the social activities in which they engage in support of the processes. Thus we see reference to the team's decisions about how they divided the activities that the task entailed, the processes involved in the task, the roles and responsibilities different team members played, and the systems used to complete the task. In a similar vein, from the extract we can visualise the meetings, communications, and negotiations that the team members needed to engage in to produce a final draft of the design report.

Another finding that emerged from examining disciplinary writing as social practice was that writing serves both to reflect and reaffirm the views on knowledge and ideologies of engineering as a discipline, which supports previous research findings (e.g. Latour & Woolgar 1986). As shown in Figure 5.1, scientific knowledge in engineering, mainly based on the basic sciences, is implemented in systemic ways to design concrete solutions to design problems. The application of theories from basic sciences to generate design solutions was clearly manifested in the student design reports we analysed. It was evident in the texts created by students and negotiated with their tutors over a number of tutorials that, to the engineering students, design represented a creative way of applying scientific principles in an attempt to solve real-world problems. For example, the titles of some of the design reports written by the students shown in Table 2 provide insight into the problems students were asked to solve in the scenarios they were provided. Designing viable solutions to such problems required them to apply scientific knowledge by synthesising existing theories and principles in order to produce flow sheets and calculations, as well as to design diagrams that graphically represented their approaches to problem-solving. As one of the lecturers mentioned about the role of supporting evidence for design reports in an interview:

> An engineer will pick the flow sheet, pick the calculations and they will tend to work with that, they know that they are the result of existing principles and theories that form the basis for designing a new project, its very fabric and that's the essential part of the project ... and then if they need to, they'll come back to the written word. (Int_02_Lec_CE)

This extract provides further evidence in support of the multifaceted epistemology of engineering which combines theories and principles of scientific, engineering, and design knowledge with skills and competences for creating numeric and graphic representations of the application of theories and principles for solving problems. It also shows the central role that flow sheets and graphics play in representing the results of the research carried out in support of professional practices (see also Curry 2014).

Table 5.2. Student design report titles

Title	School
An Ethyl Benzene Plant: Dealing with uncertainties and difficulties to improve design	Chemical Engineering
Designing a Pedestrian Operated Dump Truck for Diverse Environments	Mechanical, Materials and Manufacturing Engineering
Rehabilitation Centre: A feasibility design study	Civil Engineering
Renewable energy and electric/hybrid vehicles	Electrical and Electronic Engineering

How writing reflects and reaffirms the knowledge base (i.e. principles and theories from basic sciences) and professional ideology (e.g. from the whole to the parts) of engineering can also be seen in the talk-around-text sessions where members of the writing teams explained the processes involved in writing a design report to solve an existing problem. One example appears in the following extract from a talk-around-text session with a student writer:

> If you look at this section on alternative design … it was actually broken down which is not written here of course but we went away and did our own research to start with, you know, we looked into existing theories and projects as you do in engineering, and looked at the feasibility of some previous projects, and then came back with different designs. We then decided to cut out any redundancy, any repeats, any plant design to one, possibly two, to arrive at that … because it was theoretical, etc. We put our footnotes in before we started considering a final design that would provide a better alternative or solution to the problem … (TAT_06_3Ms)

The extract shows processes for synthesising existing scientific knowledge and previous research as a starting point for designing a solution to a problem. It is noteworthy that the student writer qualifies this process as standard practice in engineering: "as you do in engineering." The extract also describes other, more creative, processes associated with the design of solutions to a problem: spending considerable time on defining and framing the problem and considering factors that would help decide on "a better alternative or solution to the problem." These processes would allow them to build a 'bridge' between the problem space and the solution space (Cross 1997, p. 428). The extract also lends support to Gibson's (1998) views on the centrality of design in engineering as it encompasses three of the main processes – synthesis, creativity, and problem-solving—important in professional practices in engineering such as designing a project and writing a report on it.

To arrive at these findings, both applied linguists and engineers on the research team needed to develop a shared understanding of how the values and ideologies reflected in the students' writing (e.g. collaborative writing, systemic problem-solving) were shaped by the epistemologies of engineering: how knowledge and what is considered worth knowing is conceptualised in the discipline. In a similar vein, it was vital to develop an understanding of how epistemology, values, and ideologies together helped to shape disciplinary writing. In writing design reports, for example, students needed access to abstract knowledge, represented by the principles and theories provided by the basic sciences, to be able to produce solutions to real-world problems following the principles and theories of design sciences. In most cases, they approached the task by combining the two most prominent ideologies in contemporary engineering: holism and reductionism, as discussed above and illustrated in the talk-around-text session extract (see TAT_06_3Ms) where a student writer explained the processes involved in writing a design report to solve an existing problem.

Similarly, knowledge created by means of research into existing theories and how they would fit challenging problems, for instance, is used by students in writing design reports to support skills that are valued in future engineering professionals. One case is how knowledge resulting from research on possible locations for setting up an ethyl benzene plant, which would require feasibility studies, risk assessments, and cost analysis, could be used to make informed design decisions. This challenge is discursively realised in the following text from one of the design reports written by a team of students:

> Significant research was conducted into the location of the plant. A purpose built chemical production location with the required infrastructure and related industry (water supply suitable for cooling requirements, proximity and spare capacity of feedstocks, high pressure steam supplies) was finally chosen. The selected location was on an island in Singapore. Locating the plant on an island simplifies some of the environmental considerations, as this is an island with no local community. (Rep_03_CE)

As can be seen, the extract starts with a reference to research as support for decisions made and uses an impersonal style, illustrating how disciplinary epistemologies, values, and ideologies shape the ways that texts are constructed in terms of organisation of information and style. It also demonstrates the way that writing gains its "meanings and functions" within a specific social setting, as Bazerman and Prior (2003, p. 2) argue. In this way, knowledge, social meaning, and texts come together to represent a specific group of people (engineers), interacting (in writing) by using words and discourses they recognise, (e.g. impersonal text construction) in a particular context (an engineering faculty at a British university) to accomplish specific aims (demonstrating and creating knowledge).

Another central finding from this study relates to gains from conducting cross-disciplinary research. The joint research between applied linguists and engineering educators reported on here can help to inform our understandings of the role and nature of disciplinary writing, and these can enlighten pedagogical and research practices. By collaboratively examining professional practices such as disciplinary writing following the dimensions illustrated in Figure 5.3, we have reached a nuanced understanding of writing that goes beyond the mechanistic views offered by, for example, many courses in English for Academic Purposes.

As a cross-disciplinary team, we also developed shared views of the nature and dynamics of writing as social practice for the creation and dissemination of disciplinary knowledge. In addition, members of the team came to realize that the relationship between epistemic ideologies and values and disciplinary writing needs to be unpacked and discussed with newcomers to the disciplines – including students. This process of disambiguating and, to a certain extent, problematising disciplinary practices by making them visible rather than taking them for granted will also encourage the investigation and teaching of writing to become more context- and discipline-specific.

A number of pedagogical strategies can be developed from the results reported here. For instance, starting from questions such as "What does 'writing like an engineer' mean?", students can be asked to research possible answers which should cover both the linguistic and disciplinary attributes of writing in engineering. Following the discussion of students' findings, copies of successful and unsuccessful texts produced by previous cohorts of student writers can be distributed for students to evaluate and critique along a number of criteria decided on collaboratively by applied linguists and engineering educators. These criteria include how closely the texts represent the epistemologies, values, and discourses of the discipline and how well they fit the context of production. Finally, in mixed specialisation groups, students can be asked to identify and establish possible linguistic and disciplinary variations across the schools of the faculty. This exercise might help make them aware of the fact that, rather than being a generic skill, writing is a context-sensitive practice (Gimenez 2012; Jones 2009).

Conclusion

The cross-disciplinary study of writing as situated practice reported on in this chapter aimed at examining the nature and dynamics of academic and professional writing at the faculty of engineering in a university in the United Kingdom. The chapter presented an examination of the understandings of writing as situated practice that a team of researchers in applied linguistics and engineering gained from conducting cross-disciplinary research. The main results discussed here relate to writing both as social practice and as reflection of the values and ideologies of engineering as a discipline. They also show the benefits for both applied linguists and engineering educators of collaborative research and how the insights gained can inform their professional practices.

Further studies could implement the model of cross-disciplinary research illustrated in Figure 5.3 in other STEM disciplines so that new, discipline-specific models could be advanced. Similarly, cross-disciplinary studies between applied linguists and STEM professionals could extend some of our findings to cover specific issues in other related disciplines. New studies that investigate context-sensitive and discipline-specific approaches to researching and teaching writing in disciplines that have been under-represented in the literature (e.g. political sciences, architecture, and design) would help to expand the existing body of knowledge in cross-disciplinary writing research.

It is hoped that this chapter and the book overall will start a conversation between researchers and educators across disciplines and help to generate future cross-disciplinary studies that may inform the professional practices of new as well as established members of these disciplines.

References

Alley, M. (2000). *The craft of editing: A guide for managers, scientists and engineers.* New York, NY: Springer.

Barton, D., Hamilton, M., & Ivanic, R. (2000). *Situated literacies.* London: Routledge.

Bazerman, C. & Prior, P. (2003). *What writing does and how it does it: An introduction to analyzing texts and textual practices.* New York, NY: Routledge.

Bracewell, R.J. & Witte, S.P. (2003). Tasks, ensembles, and activity: Linkages between text production and situation of use in the workplace. *Written Communication, 20*(4), 511–559.

Brieger, N. (2011). *Collins English for business: Writing.* London: Harper Collins.

Charmaz, K. (2005). Grounded theory in the 21st century: Applications for advancing social justice studies. In N.K. Denzin, & Y.S. Lincoln (Eds.), *The Sage handbook of qualitative research* (pp. 507–535). Thousand Oaks, CA: Sage.

Charmaz, K. (2011). A constructivist grounded theory analysis of losing and regaining a valued self. In F. J. Wertz, K. Charmaz, L.M. McMullen, R. Josselson, R. Anderson, & E. McSpadden (Eds.), *Five ways of doing qualitative analysis: Phenomenological psychology, grounded theory, discourse analysis, narrative research, and intuitive inquiry* (pp. 165–204). New York, NY: The Guilford Press.

Christie, M., Ferdos, F. & Adawi, T. (2009). Cross disciplinary research in engineering and educational sciences: A Swedish case study. *Proceedings of the 37th SEFI Annual Conference,* Rotterdam, The Netherlands. Available at ⟨http://www.sefi.be/wp-content/abstracts2009/Christie.pdf⟩ (12 March 2013).

Cross, N. (1997). Descriptive models of creative design: Application to an example. *Design Studies, 18*(4), 427–440.

Curry, M.J. (2012). Transcending 'traditional academic boundaries': Designing and implementing a science communication course for science and engineering Ph.D. students. *Professional and Academic Writing, 40,* 4–7.

Curry, M.J. (2014). Graphics as invention heuristics in writing for publication by academic engineers. In M. J. Curry & D. Hanauer, (Eds.), *Language, literacy, and learning in STEM Education: Research methods and perspectives from applied linguistics* (pp. 87–106). Amsterdam: John Benjamins.

Engineers' Council for Professional Development. (1947). *Canons of ethics for engineers.* New York: Engineers' Council for Professional Development.

Figueiredo, A.D. (2008). Toward an epistemology of engineering. *Proceedings of the Workshop on Philosophy & Engineering* (pp. 94–96). London: Royal Engineering Academy.

Gibson, I.S., (1998). Assessment criteria for undergraduate project work in engineering design. *European Journal of Engineering Education, 23*(3), 389–395.

Gimenez, J. (2008). Beyond the academic essay: Discipline-specific writing in nursing and midwifery. *Journal of English for Academic Purposes, 7*(3), 151–164.

Gimenez, J. (2012). Disciplinary epistemologies, generic attributes and undergraduate academic writing in nursing and midwifery. *Higher Education, 63*(4), 401–419.

Gimenez, J. & Thondhlana, J. (2012). Collaborative writing in engineering: Perspectives from research and implications for undergraduate education. *European Journal of Engineering Education, 37*(5), 471–487.

Gygi, K. & Zachry, M. (2010). Productive tensions and the regulatory work of genres in the development of an engineering communication workshop in a transnational corporation. *Journal of Business and Technical Communication, 24*(3), 358–381.

Henry, J. (1886). *Scientific writings of Joseph Henry*. Washington, DC: The Smithsonian Institution.

Isherwood, B.F. (1863). *Experimental researches in steam engineering*. Philadelphia, PA: W. Hamilton.

Ivanic, R. (1998). *Writing and identity*. Amsterdam: John Benjamins.

Jones, A. (2009). Redisciplining generic attributes: The disciplinary context in focus. *Studies in Higher Education, 34*(1), 85–100.

Kim, H.C. & Severison Eklundh, K. (2001). Reviewing practices in collaborative writing. *Computer Supported Cooperative Work, 10*(2), 247–259.

Latour, B. & Woolgar, S. (1986). *Laboratory life: The construction of scientific facts*. Princeton, NJ: Princeton University Press.

Lillis, T. (2008). Ethnography as method, methodology, and 'deep theorizing'. Closing the gap between text and context in academic writing research. *Written Communication, 25*(3), 353–388.

Luzon, M.J. (2005). Genre analysis in technical communication. *IEEE Transactions on Professional Communication, 48*(3), 285–295.

Nelson, S. (2000). Teaching collaborative writing and peer review techniques to engineering and technology undergraduates. *Proceedings of the 30th ASEE/IEEE Frontiers in Education Conference*, S2B-1–5.

McClellan, W. (1913). A suggestion for the engineering profession. *Transactions of the American Institute of Electrical Engineers, 32*(2), 1271–72.

Parks, S. (2001). Moving from school to the workplace: Disciplinary innovation, border crossing, and the reshaping of a written genre. *Applied Linguistics, 22*(4), 405–438.

Seifert, C. (2009). Writing for business: A graduate level course in problem-solving. *Business Communication Quarterly, 72*(2), 200–213.

Sherwin, J. (2010). An approach towards holism in science and engineering. *Journal of Systems Science and Systems Engineering, 19*(3), 285–305.

Storch, N. & Tapper, J. (2009). The impact of an EAP course on postgraduate writing. *Journal of English for Academic Purposes, 8*(3), 207–223.

Strauss, A.L. (1987). *Qualitative analysis for social scientists*. Cambridge: Cambridge University Press.

Street, B.V. (2001). *Literacy and development: Ethnographic perspectives.* London: Routledge.
Tenopir, C., & King, D.K. (2004). *Communication patterns of engineers.* New York, NY: John Wiley & Sons.
van Emden, J. (2005). *Writing for engineers.* Houndmills, UK: Palgrave.
Wang, X. (2008). *Software engineering foundations.* New York, NY: Auerbach Publications.
Wheeler, E. & McDonald, R. (2000). Writing in engineering courses. *Journal of Engineering Education, 89*(4), 481–486.
Winsor, D. (1996). *Writing like an engineer: A rhetorical education.* Mahwah, NJ: Lawrence Erlbaum Associates.
Winsor, D. (2001). Learning to do knowledge work in systems of distributed cognition. *Journal of Business and Technical Communication, 15*(1), 5–28.
Zhu, W. (2004). Writing in business courses: An analysis of assignment types, their characteristics, and required skills. *English for Specific Purposes, 23*(2), 111–135.

CHAPTER 6

Graphics and invention in academic engineers' writing for publication

Mary Jane Curry
University of Rochester

This chapter explores how academic engineers write for publication, focusing on "invention" – that is, moments when writers identify the research results they want to present and decide on the arguments they want to make in an article. A key finding presented is that beyond the well-documented role of graphics in displaying research results, graphics also play a crucial heuristic role in invention. This finding emerged from an ethnographic study of three engineering research groups, which entailed the analysis of a range of qualitative data to offer perspectives on the experiences of academic engineers writing for publication. Drawing on this research, the chapter documents that in developing texts for publication, engineers often begin with the graphic results of data analysis to identify findings and begin to craft arguments. Further, in research group and informal meetings, engineers invoke the notion of storytelling through graphics as they socialize their group members into the practices of research dissemination via posters and articles.

Introduction

A chief feature of communications practices in the field of engineering is multimodality, as Mathison (2000) notes: "an engineer's way of knowing involves multiple symbol systems, some of which are verbal and some of which are numerical and visual" (p. 75; see also Amare & Manning 2007; Archer 2006; Selzer 1983). Research has identified the recursive interplay between text and graphics in the communication of engineering knowledge by undergraduate students (Archer 2006) and engineers in the workplace (Hutto 2007; Lloyd 2000). In presenting new knowledge, "visual representations of data are the work horses of argument," as Poe, Lerner, and Craig (2010, p. 115) claim in their overview of graphics and the scientific research article. Graphics therefore appear to be integral to the construction of knowledge and its communication across a range of contexts in which engineering texts are produced. This chapter focuses on the heuristic role that graphics play in invention in the professional writing practices of academic engineers. Addressing the need for more empirical understandings of how scientists and engineers write, it presents findings from a three-year ethnographic study of the research dissemination practices of academic engineers (faculty members and their research group members), as they write professional genres such as conference papers/posters and journal articles. Ethnographic research can provide a window onto engineers' actual writing practices, enabling us to document practices and the perspectives of writers. Because it adopts an insiders' or "emic" perspective on particular social practices, ethnographic research enables an analysis of engineers' writing practices in terms of their meaning for both writers and the construction of disciplinary knowledge.

In this chapter I argue that (a) engineers often envision and construct the argument of a paper through the graphics (visuals such as tables, figures, and plots) that they generate in data analysis – thus in their writing visuals play a more central role than simply as illustrations for previously determined research findings and arguments; (b) the work of constructing findings and articulating arguments in text is a social, rather than an individual, practice; and (c) engineering faculty members use the metaphor of storytelling to conceptualize the development of a paper's argument and to persuade their readers. Faculty members draw on this storytelling metaphor to advise post-doctoral fellows, graduate students, undergraduates, and technicians working on specific projects as well as writing more generally.

The chapter first reviews the literature on multimodality and academic writing, then discusses the theoretical framework of academic writing as a predominantly social rather than individual practice (Curry 2003; Lea & Street 2006; Lillis & Scott 2007). I link this theoretical stance to the social turn in understandings of the rhetorical notion of invention (Lauer 2004; LaFevre 1987). After describing the three-year ethnographic study of engineering research groups, I present my findings about graphics and invention. I conclude with a discussion of the value of ethnographic research for understanding what goes on in the professional communication practices of engineering faculty members and students. These findings have implications for teaching and advising students and early career scholars about crucial communication practices in engineering, and by extension, other STEM fields.

Theoretical framework: Engineering writing as multimodal social practice

Social practices are repeated patterns of activity aimed at particular purposes, such as, here, publishing and presenting academic research (Wenger 1998). A social practice perspective takes account of the range of activities comprising a practice and considers how people learn the practice. It posits that learning occurs through increasing participation in valued practices, often under the guidance of more expert members (Barton & Tusting 2005). A social practice perspective helps with understanding how students, for example, gain access to academic knowledge-making practices (Lillis 2001). In the present study, learning academic publishing practices involves interactions among students, post-doctoral fellows, and faculty research group leaders. It illustrates what Lea and Street (2006) describe as an academic socialization model, in which learners are seen to participate in "the ways of talking, writing, thinking, and using literacy that typify members of a disciplinary or subject area community" (p. 368).

Scholars concerned with learning about the communication practices of academic disciplines have focused on novices producing writing that meets specific genre expectations in educational and workplace settings (e.g. Li 2002, 2005; Poe et al. 2010; Winsor 1992). In the field of composition studies, a predominant focus on the individual student has reinforced the notion that invention is a stage of writing that takes place before writing (as in 'pre-writing' in the process approach to writing) as well as being an individual practice (Bawarshi 2003). Invention, according to a summary by Odell and Swersey (2003, p. 40), was "for Aristotle, the discovery of persuasive arguments; for modern rhetoric, the formulation and articulation of ideas" – or the "discovery and creation models" of invention (Hawhee 2002, p. 17). Instead, Hawhee proposes the notion of "invention-in-the-middle," which connects to social practice theories by acknowledging that generating ideas and developing arguments occur in social encounters. These encounters include the collaborative writing that characterizes engineering (Gimenez & Thondhlana 2012; Winsor 1994) and indeed encounters taking place in an "individual" writer's head, as writers are always knowingly or unknowingly responding to the voices of others (Bakhtin 1986; LaFevre 1987; Lillis 2003).

Problematizing notions of the nature of engineering writing itself, Winsor (1992) argues that research data per se – represented as tables, figures, charts, and other graphics – in fact constitute a form of writing. In Winsor's study, "the engineer actually worked very hard to make these written traces [of research activity] appear" in his text (p. 342). She advocates that researchers move beyond "the insistence on the presence of words [that] comes from the deeply ingrained idea that writing is recorded speech" (p. 342; see also Winsor 1994). This perspective aligns with research pinpointing academic knowledge as increasingly being constructed and presented through a range of modalities (Bezemer & Kress 2008; Hanauer 2006). For example, in ethnographically studying the "linguistic landscape" of a microbiology laboratory, Hanauer (2009) documents how scientists and students graphically represented the knowledge they were constructing in the laboratory and used such representations (e.g. conference posters) as learning tools. Likewise,

Dicks, Flewitt, Lancaster and Pahl (2011) argue for the value of ethnography in studying multimodality, claiming that "any instance of literacy is in fact a multimodal 'event'" (p. 230). Nevertheless, considerations of multimodality have often been downplayed in relation to the written word, as I discuss next.

Research on writing for publication in science and engineering

While some of the literature on scientists writing for publication (e.g. Bazerman 1988; Knorr-Cetina 1981; Latour & Woolgar 1979) makes passing note of the graphic display of research findings, most researchers do not closely examine its role.[1] However, in a rare study of the interactions among scientists as they write for publication, Rymer (1988) documents how nine "esteemed" biochemists write articles. Among her findings, she points to the generative role of graphics:

> Some invention procedures – using graphs, tables of data, and the author's own previous papers as planning devices – characterize all the [scientists'] practices. Typically, figures and tables of the data ... function as visual aids to invention and as organizing points during planning and drafting. (p. 220)

In contrast to Rymer's apparent assumption that graphical representations of data have a function in invention, the literature on scientists learning to write for publication concentrates on the social practices of writing. For instance, Blakeslee's (1997) study of collaboration among a physics graduate student, post-doctoral fellow, and faculty member in a research group describes the student drafting his first article. Although she discusses the visual display of information, it is only mentioned as embedded in a data extract about how the student approaches the presentation of "information appropriate for their auditors." The student recounts:

> When I say what kind of results, I mean what kind of figures – like all these tables I have to get rid of [in the revision]. Now I have to decide what to put instead. I'll base my decision on results on which ones actually show that one method is better than another. (p. 146)

Here again, however, graphics receive only fleeting attention. In engineering specifically, the literature on writing for publication has mainly analyzed published texts and genres (e.g. Koutsantoni 2006; Luzon 2005; Shehzad 2006) or described and evaluated workshops and other instructional activities (Alford & Stubblefield

1. In fact, none of the 40 studies that Swales (1990) includes in Table 3, 'Overview of the Textual Studies of the English RA [Research Article]', focuses on the role of graphics; article structure is the primary concern of many researchers.

2002; Leydens & Olds 2007). Of the few investigations of professional engineers' writing practices, Shaw's (2010) study of the research report as deployed in the distinct but related domains of academia and industry demonstrates the relationship of context, purpose, and expertise in writing for different audiences. In a recent study of engineers writing for publication, Chiu (2011) focuses on two early career Taiwanese scholars. One of them, a mechanical engineer, Eric, "wrote the first draft only after organizing experiment data and discussing the major findings with his advisor and lab mates. ... Eric negotiated with his data and tried hard to find a perspective by which to present his claims" (p. 472). Perhaps in foregrounding scholars' use of English as an additional language, this article presents findings about their "extensive reading and textual borrowing strategies" (p. 472) but does not explore the role of graphics. Overall, then, much of the literature on writing science and engineering journal articles mentions graphics only incidentally, with little sustained exploration of its role in writing. However, as Graves (2005) cautions in discussing scholarly publications:

> the process of creating the knowledge presented in academic articles is complex and multifaceted, suggesting that conclusions about a particular academic discipline that use as their evidence only the product – the written and published article – do not represent the whole story especially if claims about process are extrapolated from the written text. (p. 255)

The paucity of research on the practices of academic engineers writing for publication was one prompt for the present study, which I turn to next.

Methodology of the larger study

This paper comes from a larger study, "Publishing Engineering Writing" (PEW), whose primary research question was: "In what experiences or activities do engineering faculty members, post-doctoral research fellows, and students engage in writing for research publication?" In this chapter I address two sub-questions:

- What are the professional writing practices of academic engineers?
- How do engineering students and post-doctoral research fellows learn the practices of writing for publication?

Ethnographic methodology offers an approach to understanding the lived experiences of writers in terms of their practices over time (Kress 2011; Lillis 2008). Ethnographic methods of data collection include participant observation of specific contexts; video- and audio-taping of specific events; conducting focus groups; and holding individual interviews. To study the practices of writing for publication, the PEW study adopts the "text-ethnographic" methodology developed to explore writing for publication within specific contexts of production (Lillis & Curry 2006, 2010). Text-ethnographic methods include the collection of texts written for publication as well as other relevant documents including participants' correspondence with journal gatekeepers. Analytic methods used in text-ethnographic methodology follow the principles of modified grounded theory (Charmaz 2006). This approach involves multiple readings of interview/meeting transcripts, field notes, and documentary data in order to create codes, categories and themes that signal specific research findings. Researchers then validate the analytic findings partly by means of member checking, that is, presenting preliminary findings to research participants for feedback.

Study setting and participants

The PEW study took place over three years at a U.S. engineering school with 81 tenure-track faculty members. The three faculty members and research groups that participated were representative of the school in terms of gender, race, and national/ethnic origin. However, they came from only two sub-disciplines, albeit large departments: electrical and computing engineering and biomedical engineering. Table 6.1 shows the composition of each research group (names are pseudonyms).

Table 6.1. Participants in the PEW study research groups

Research group	Head	Members	Gender	Countries of origin
A	Professor Arthur, a white, male, distinguished full professor and department chair	13 students and post-doctoral research fellows	3 women 10 men	United States, East Asia, South Asia, central Europe
B	Professor Brown, a white, male, full professor and former department chair	six students and post-doctoral fellows	1 woman 5 men	United States, East Asia, central Europe
C	Professor Courts, a white, female associate professor (also collaborating with two faculty colleagues, one at this university and one at another institution)	six students, one post-doctoral research fellow, three technical staff members	5 women 5 men	United States, East Asia, South Asia

Data collection consisted of audio- and video-recorded focus groups with various sub-groups (female faculty, female undergraduates, users of English as an additional language); 47 audio- and video-recorded individual semi-structured interviews with the participants; more than 100 audio and/or video recorded participant observations, with attendant field notes made of formal and informal meetings including Professor Courts's weekly meeting; and multiple drafts of more than 50 texts written for presentation and publication.[2] Interviews and selected meeting recordings were transcribed and data were analyzed using modified grounded theory, as described above. Various data sources were transcribed and coded by me and members of my research team over the three years of data collection and two subsequent years of analysis. Using emerging codes and themes, data were coded independently by research assistants. Resulting analyses were compared, then codes and themes adjusted and reapplied to coded and new data. Ethnography enables us to triangulate emerging understandings of participants' practices by drawing on complementary data sources. In this case, triangulation allowed us to compare participants' self-reports with descriptions of these practices from their collaborators and the researcher team's documentation of practices. Drafts of this paper have been shared with participants, who confirmed our analyses of their experiences and practices.

Findings: Graphics as invention in writing for publication

As mentioned, two key and related findings from the ethnographic study are that (a) graphics function as invention heuristics in the professional writing of academic engineers and (b) academic engineers convey these practices to those working with them either implicitly through shared practices or explicitly in discussion, often invoking the notion of storytelling. In this section, I use extracts from individual and focus group interviews to construct an overview of the faculty members' practices in directing their research groups' writing activities. Then, drawing on observational data and recordings as well as interview data, I describe a meeting that Professor Courts held in order to begin conceptualizing a paper.

As noted, the three engineering faculty members were highly successful, conducting and publishing research in their sub-disciplines that was funded by large grants and supporting a number of Ph.D. students and post-doctoral fellows. An important writing and teaching practice shared by the engineering faculty members was to assign to group members the initial drafting of papers and conference posters, to which the faculty members then responded. However, the majority of

2. My gratitude to research assistants Hairong Shang, Hee-Jeong Oh, Nan Zhang, Rachel Chaffee, Farzana Hafsa, and Qiao Li for support in data collection, management, and analysis over the life of the project. Thanks also go to my doctoral student group, which read an earlier draft of this paper, as well as to Julio Gimenez, David Hanauer, and Fredricka Stoller for their suggestions. Earlier versions of this paper were presented at conferences of the Canadian Association of Applied Linguistics (2009) and the Symposium on Second Language Writing (2010).

their other practices for engaging group members in preparing conference posters and writing articles differed, which affected my own research team's access to observing their practices. For example, Professor Arthur rarely held meetings with his entire group, which was the largest of the participating groups; rather, he moved multiple projects forward simultaneously in small and often quickly scheduled meetings, impromptu hallway conversations, and through email. Professor Brown generally followed similar practices. Thus it was difficult to be on site when interactions related to writing took place in these professors' groups. In contrast, in the second year of the study, Professor Courts initiated an ongoing weekly writing meeting with graduate and undergraduate students, a post-doctoral fellow, and technical staff members and invited me to research the group's activities. Although Professor Courts also met outside of scheduled meetings with smaller groups or individuals working on specific papers, these smaller groups would also bring drafts of these papers to the whole group for peer review and discussion. We were able to observe these scheduled meetings because Professor Courts diligently notified us about them and included me on emails to her group and sub-groups.

In terms of this chapter's focus on graphics, the importance of graphics in the faculty members' writing practices emerged as a strong theme in the data analysis, as illustrated in comments made by Professor Arthur:

> *I like pictures and figures. ... You put the text after the figures.* This is ... actually how we write the paper. *We visualize the set of pictures, the story that we tell, and then we put the text around it.* That's how ... [but it] never works the other way around. Because especially if you write a short paper, you have to get to the point immediately and *you have to tell a story*, because people have no patience in my field. (Interview; emphasis added)

Professor Arthur's comments point to the role of graphics as invention heuristics and in structuring the argument of the paper, which he calls "the story." These comments also highlight his well-developed sense of his audience as busy colleagues – "people [who] have no patience" – and whose lack of time adds pressure for publications to get to the point quickly. Students in Professor Arthur's group corroborated his description of these practices. Yuan, an international Ph.D. student, explained their practice:

> What we [students] do is discuss with [Professor Arthur] about the possibility of writing a paper. *I will show him those graphs*, those results I have, some experimental results, some theoretical results. And he will comment on it. And then he will say, 'Okay, this is a good paper. You should write it.' (Interview; emphasis added)

Yuan's comments underscore how the visual presentation of research results ("those graphs") contributes to the research group determines whether such results might be publishable.

Turning to Professor Brown, his summary of his typical way of working with a research group member on a paper follows the stages suggested by many writing teachers – outline, draft, revision:

> Generally in the meetings with the student, when we mutually agree that there is something worth writing up, then I will ask them to write an outline first. We will sit down with the outline and then we edit that, and then figure out the content, the flow of the paper. And then I will make them write a first draft and then work on that. I try to get them to do all the writing. I make small changes, but it varies. We will discuss the paper and the structure, what would be in it. (Interview)

Any mention of the role of graphics in identifying "something worth writing up" seems notably absent from Professor Brown's description of the writing process he follows and promulgates. However, data from his group members reveals that the advice he imparts to them – and what they actually do – is to begin writing a paper by creating and contemplating visuals that display research results. For instance, Natalia, an international post-doctoral fellow, believes that Professor Brown:

> has a better [approach than mine], I tried it a couple of times and you just jot down figures and then you write the captions and describe what's on the figure like in a text.... To write a paper, that's the process where you put those figures and you describe what you did and you put in the references. (Interview)

Likewise, one of Professor Brown's international Ph.D. students, Christopher, reports the advice he received to pursue these writing practices:

> When we find results that are pretty interesting, we look at the results and work in a kind of backwards way to say, we will put the interesting things that we have discovered and then back it up, support with our findings and our data. One interesting thing Professor Brown said when we were first writing together, he said, "Don't think of how to write it, just put the pictures, images that you think are interesting and then write about it." (Focus group interview)

Thus in practice, Professor Brown shares Professor Arthur's perspective on how graphics function as invention heuristics in writing; the fact that he does not explicitly identify how graphics function in his writing and supporting research group members may reflect a conventional sense of what writing practices consist of, as Winsor (1992), discussed above, identifies. The data extracts included in this section point to the nuanced understandings that research group members have gained about the practices and advice given by professors Arthur and Brown.

In discussing her writing practices, rather than summarize them as did her colleagues, Professor Courts articulated that she and her group members follow a routine of beginning an academic paper by creating an outline and using it to support student writing. Nonetheless, she lamented:

> some of the students don't want to outline anything... Some of them just want to start writing. Or the first [draft] they give you they've really written it all out and so it is harder. But I have [an international] post-doc right now and his language skill is not very strong. So I really force him to do outlines. So we start from scratch.

Although Professor Courts's comments suggest that she follows the process approach to writing in which outlining is a pre-writing technique, the ethnographic data show that the outlines created in her research group in fact involve graphical representations of research results. As her post-doctoral research fellow, Ali, recalled:

> Usually we have to make an outline, what to include and what not ... The first paper that I actually wrote in this lab, *you make an outline what kind of figures we should show, or just the results section ...* what are the key figures they should show. Once you know these are the results you want to show, then you can come back to the introduction. (Interview, emphasis added)

The term "outline," therefore, suggests a type of multimodality that writing teachers – and academic engineers who are not writing specialists – may not recognize. This possibility is not surprising given that tacit knowledge about a social practice is not always available for expert practitioners to articulate, whether to their students or researchers (Nathan & Petrosino 2003). To explore these practices in more depth, I next describe the use of the analytic method of "text history" (Lillis & Curry 2006) to present a fine-grained rendering of the function of graphics as invention heuristics. Text histories draw on the range of qualitative data listed above to trace the development of a particular text over time, an approach particularly suited to the long time frames involved in the development of academic texts for publication.

Text history: Launching a research article

This partial[3] text history[4] is based on recorded participant observations, field notes, PowerPoint presentation slides, and one draft of an article for publication. I draw on these data to describe how Professor Courts and her group began to

3. This text history is partial because Professor Courts and her collaborators put this paper on hold to write other papers during the data collection period.

4. To protect the anonymity of participants in the study, details of the data that might disclose their research areas are omitted and replaced with ellipses or [X], [Y], etc.

conceptualize an article in a one-hour meeting with her post-doctoral fellow, Ali, one of her technicians, Kathy, and an undergraduate, Nathan. Before the meeting began, Professor Courts had identified the target journal for the article, which she considers the top journal in her field, and had provisionally decided that the results the group was about to discuss would be publishable in this journal.

Professor Courts begins the meeting by noting, "I've got a lot of figures and things and I looked up the guidelines in [the journal] to see if we're going to shoot for that [journal], which seems worth doing." She then reads aloud an extract from the journal's website, which she had pasted into the draft document:

> **An article** is a substantial novel research study, with a complex story often involving several techniques or approaches. The main text (excluding abstract, Methods, references and figure legends) is 2,000–4,000 words. … Articles have no more than 8 display items (figures and/or tables). An introduction (without heading) is followed by sections headed Results, Discussion and Methods. … (Extract from Paper Draft 1, boldface emphasis original)
>
> Addressing the group, Professor Courts comments,
>
> I was just starting to remember the results we've got and of course some of them we're still pulling together the figures for. But we could map out what figures we want to include. I think that's a good way to start. One thing I just did was to run the [X] analysis program, but I think it's easier to look at this PowerPoint™ from a talk I gave in Munich last year.

She then begins to project the 25 slides of her conference presentation, of which 13 either included or entirely comprised visuals. This move itself illustrates the multimodal nature of the process of invention in this conversation: She is using a Microsoft Word™ document in for the article draft and slides from a PowerPoint™ presentation which include the data output in the form of graphs. Professor Courts continues:

> This is just to remind us what figures we've got. And *the general story to remember* is that [summary of the research results from a former Ph.D. student in her group that underpin this study]. … Most papers in the literature would argue that [X] and you can argue that [Y] … that's what we've got to shore up, is the [Y] analysis. … So as we go through these figures it is useful to see, one figure might be the apparatus, I don't know how critical it is to have that in this paper. (Recorded observation transcript; emphasis added)

Ali, the post-doc, then points out that because they used a commercially produced apparatus in doing the research, a description should already be publically available. Kathy, the technician, wonders if Professor Courts's laboratory had previously published a paper with a description of the apparatus, to which the present paper could refer.

> Professor Courts: Not exactly. But it's an optional figure, it might make it easier for people to see. A picture is worth a kilo-word, so if we include it, it would make the Methods longer, but the Methods are not included in the word count, right? It makes the Methods longer if you don't have the figure so I guess it's just a trade. (Recorded observation)

Here Professor Courts points to the tension created between the requirements of the target journal – limiting authors to eight "display items (figures/tables)" and word limits for certain sections of the article – and the need for rigor in the Methods section that would be established in part by describing the apparatus – whether in words or visuals. As the discussion moves on to which types of results to include from experiments with two types of subjects, Professor Courts advocates including both sets:

> Professor Courts: It fits better into this mold of multiple techniques.
> Ali: *Complex story?*
> Professor Courts: [*chuckling*] *Complex story.* (Recorded observation, emphasis added)

While neither of these participants elaborates here, Professor Courts's laugh suggests that the story metaphor that Ali introduces is part of an ongoing conversation in her group (which my ethnographic data verify). Next she presents the slide shown in Figure 6.1, saying, "This is one of the results for [X]."

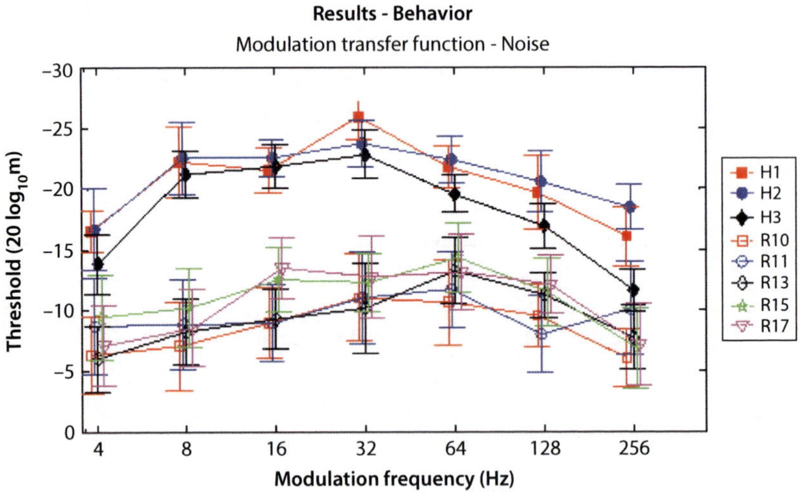

Figure 6.1. Experimental results from Professor Courts's research group

The group embarks on a cost-benefit analysis of paying the journal to include the graphic in color. Here the main considerations are the pragmatic in terms of cost and how color might be reproduced in photocopies or computer print-offs of an article. Of interest here is how important the subject of the graphics in the proposed article remains, in terms of the group's shared understanding of the value that appropriately presented graphics add to the paper even as future reproductions.

When Professor Courts projects the second results slide, a discussion ensues about whether to include the down arrows (visible just above 4, 8, and 16 on the x-axis in Figure 6.2) that represent subjects' performance on an experiment. When Kathy queries whether it is necessary to include the down arrows, Ali suggests that the figure legend could explain their significance. Professor Courts proposes writing a description of the results and their significance and removing the down arrows, then immediately counters this possibility, saying that removing the down arrows might cause readers to miss the point. Here again what might appear to be minutiae in terms of choices about visual presentation take on significance in relation to the argument to be made and how readers might receive it. Professor Courts mentions relevant results from a recently published article, concluding, "Maybe it's worth playing around with the down arrows."

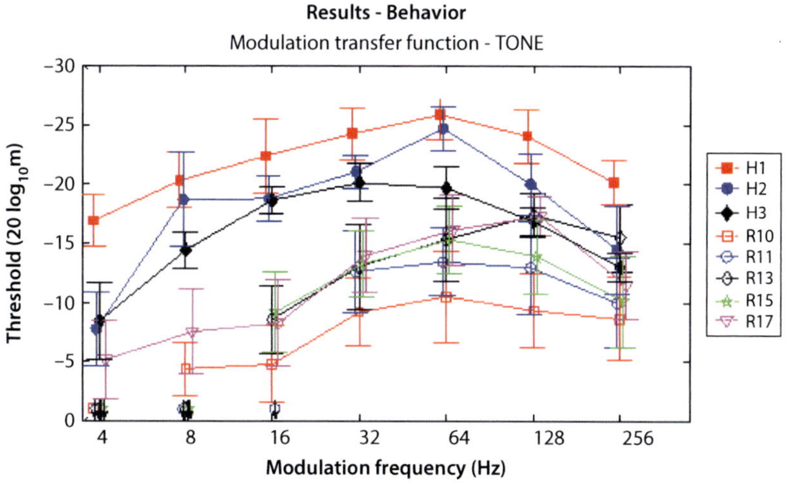

Figure 6.2. Experimental results using down arrows

These data extracts and illustrative figures show how a five-minute conversation about what might seem a trivial notation – the down arrow – prompts the group's discussion about presenting a key argument in the paper. In addition, in mentioning a related recent article, Professor Courts steers the group's thinking about research results such that they craft their argument to contribute to the "conversations of the discipline" (Bazerman 1980; see also Shaw 2010).

Invention in relation to writing an engineering article for publication therefore involves considerations of audience (here both the specific target journal and the end-user who will interact with color or black-and-white versions of the published article), how the graphics are constructed (decisions about whether to include the down arrows in Figure 6.2), and the larger research conversations taking place.

As the meeting continues, Professor Courts recaps the functions of Figures 6.1 and 6.2: "These two plots summarize all the behavior [data]. So at least it's concise." When she shows her subsequent PowerPoint slides, Nathan suggests combining the slides shown in Figures 6.3 and 6.4 to introduce the second set of data from which the results come.

Contemplating the results displayed on these two slides, Professor Courts comments,

> *This story* is a little different than the poster, which was earlier. In the poster, the [X] thresholds weren't good and maybe we were goofing somehow when we were calculating because when we went back last summer they were fine. … We'll repeat all these. These are the plots that we need to do, the newly clustered data. *So this story is pretty simple.* It's just a matter of describing. Now … we're only showing [X]; there's a whole lot that fall off the bottom. We have to describe the fact that that's true and actually give numbers. (Recorded observation; emphasis added)

Here the metaphor of story is interwoven in the discussion about articulating the paper's argument about the results. As the meeting continues, Professor Courts records the group's ideas for the paper on the word-processed document on her laptop. Although space does not permit a full explication of the hour-long meeting, these examples show the central role of graphics in the conceptualization of this article: deciding which results to argue for, how many visuals would be needed to convey the argument (the results), and the specific form the graphics should take. This text history also documents how identifying research results can take place over long spans of time, as group members considered the slides that Professor Courts presented a year earlier to be provisional in the process of writing the article and assumed that they would need to revisit and refine their experimental and communication tasks. Indeed, she mentions that the slides showed slightly different results than did an even earlier poster presenting research results. The text history also reinforces understandings of the collaborative nature of writing for publication in STEM, a process beginning with the collaborative invention of research results through the heuristic of the data graphics.

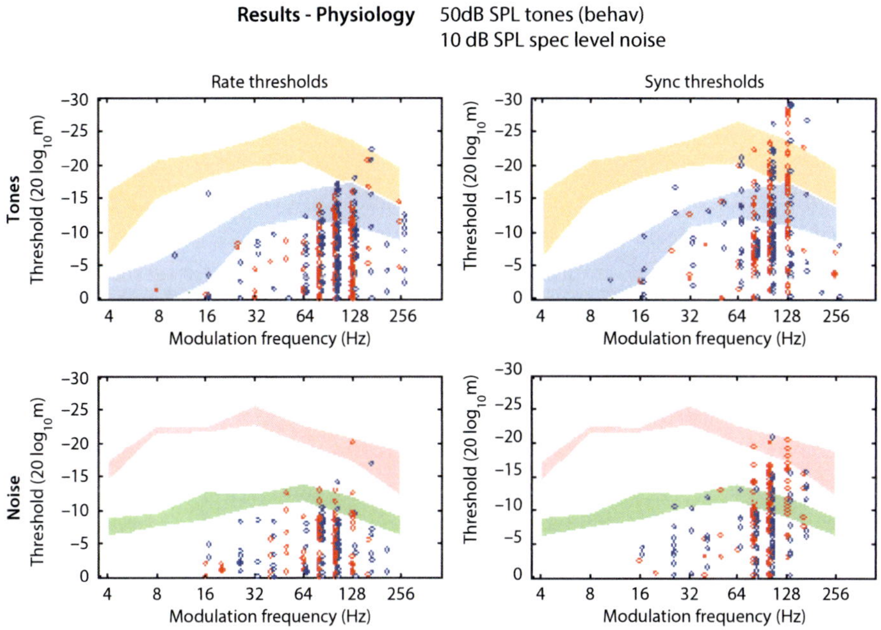

Figure 6.3. Experimental results with one set of subjects

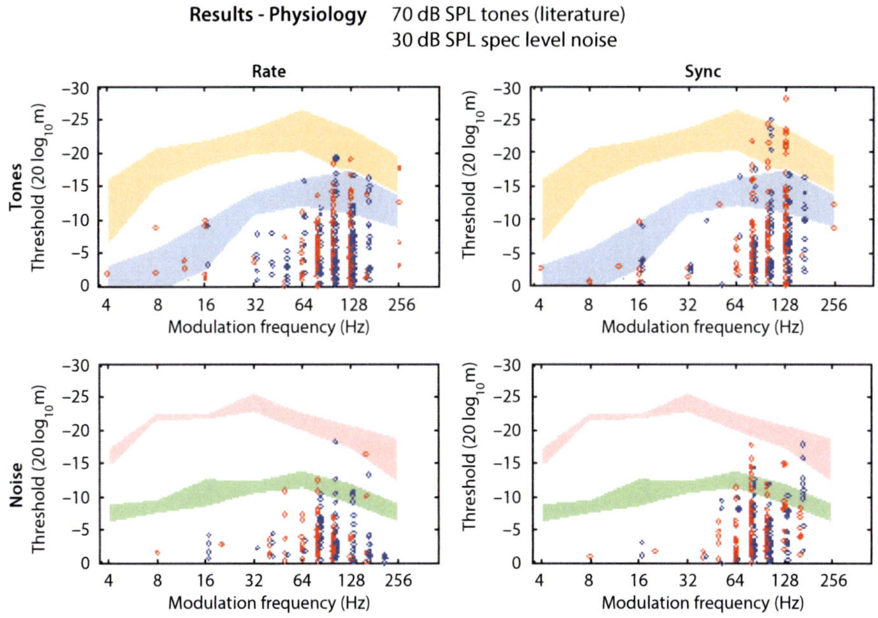

Figure 6.4. Experimental results with a second set of subjects

Discussion and implications

This chapter has documented the heuristic role of graphics in the writing of academic engineering research papers; in contrast to having merely ancillary and illustrative functions, graphics act as invention devices that enable engineers to map out arguments to put forward in research dissemination, often conceptualized as a "story." Participants both articulated and enacted these practices in the three years during which the ethnographic study took place. Ethnographic research allowed the documentation of a range of practices and participants' perspectives in order to arrive at these findings. They suggest that academic engineers do not reserve their consideration of the graphical display of research results for the final stages of writing, but rather begin with the visuals. These findings lend support to Winsor's (1992, 1994) argument for the revision of conceptions of scientific/technical writing that divorce text and visuals. Indeed, both engineers' writing processes and their informal teaching practices manifest a clearly multimodal nature. In various ways, the faculty members in this study articulated or enacted this understanding with the members of their research groups and with smaller teams writing research publications – even if they did not include them in their descriptions of their writing practices.

These findings have implications for understanding and sharing the social practices of writing for publication in academic engineering, and by extension, in other STEM disciplines. For writing instructors as well as disciplinary faculty members working with students on writing projects, these findings suggest the value of considering the role of graphics in text production not only as illustrations but also as heuristics for invention in identifying research findings and shaping rhetorical arguments. Along these lines, Poe et al. (2010) provide an example of an MIT faculty member, Dennis Freeman, who teaches physiology students to use a storyboarding technique that includes these three concepts:

> 1. Data driven scientific research. Organize and locate trends in data before beginning to write the supporting text. 2. Each figure in a report tells its own story. Design figures that make the point that you want to make. 3. In sum, the figures in a report tell the narrative of the research. Consider if the data make a logical sequence from one figure to the next figure. (p. 118)

Based on the literature, Freeman's example represents a rare articulation of such a strategy, but one that aligns well with the findings presented here.

Another implication of this study is to support the suggestion that writing guides and other materials on research dissemination should be grounded in empirical evidence about engineers' actual writing practices (Curry & Lillis 2010, 2013; Harwood 2005). Such guides tend to work from an assumption of an individual author, to privilege the production of written text over multimodality, and to advocate a linear writing process that relegates the function of graphics to illustrating points made in the text.[5] In fact, a comment by one of Professor Brown's students, Osman, supports this point: "I sometimes feel that the books telling how to write, it's not really based on the practical writing, it's just in theory" (Interview). Others have noted the mismatch between advice guides/writing textbooks and the empirically documented practices of scientists. Indeed, Poe et al. (2010) consider that "part of students' misconceptions about the function of visual representation comes from textbooks" (p. 116).

In addition, this study suggests that research on engineering and scientific writing should widen its lens to include a broader focus on multimodality. It seems imperative for an intrinsically multimodal discipline such as engineering to account fully for the uses of graphics in writing. Finally, the metaphor of story and storytelling embedded in the participants' comments about graphics deserves further attention; I am analyzing additional PEW study data for the ways that engineers use "story" as part of their writing and teaching about writing for publication.

References

Alford, E. & Stubblefield, P. (2002). Mentoring engineering graduate students in professional communications: An interdisciplinary workshop approach. American Society for Engineering Education *Annual Conference Proceedings,* Montreal, Quebec, 11835–11849.

Amare, N., & Manning, A. (2007). The language of visuals: Text + graphics = Visual rhetoric tutorial. *IEEE Transactions on Professional Communication, 50*(1), 57–70.

Archer, A.H. (2006). A multimodal approach to academic 'literacies': Problematizing the visual/verbal divide. *Language and Education, 20*(6), 449–462.

Bakhtin, M. (1986). The problem of speech genres. In C. Emerson & M. Holquist (Eds.), *Speech genres and other late essays* (trans. V.W. McGee). Austin, TX: University of Texas Press.

Barton, D. & Tusting, K. (2005). (Eds.) *Beyond communities of practice: Language, power, and social context.* Cambridge: Cambridge University Press.

Bawarshi, A. (2003). *Genre and the invention of the writer: Reconsidering the place of invention in composition.* Logan, UT: Utah State University Press.

Bazerman, C. (1980). A relationship between reading and writing: The conversational model. *College English, 41,* 656–61.

5. As Bennett (2009) attests, the marketplace is replete with guides to writing and publishing; see Curry (2011) for a review of five general guides to academic publishing. Kamler and Thomson (2008) critique the panoply of dissertation writing guides for doctoral students. A key finding from their analysis is that such guides "package their contents as a series of defined, linear steps" (p. 510).

Bazerman, C. (1988). *Shaping written knowledge: The genre and activity of the experimental article in science.* Madison, WI: The University of Wisconsin Press.
Bennett, K. (2009). English academic style manuals: A survey. *Journal of English for Academic Purposes, 8,* 43–54.
Bezemer, J. & Kress, G. (2008). Writing in multimodal texts: A social semiotic account of designs for learning. *Written Communication, 25*(2), 166–195.
Blakeslee, A. (1997). Activity, context, interaction, and authority: Learning to write scientific papers in situ. *Journal of Business and Technical Communication, 11*(2), 125–169.
Charmaz, K. (2006). *Constructing grounded theory: A practical guide through qualitative analysis.* Thousand Oaks, CA: Sage.
Chiu, Y.-H. (2011). Exploring non-native science scholars' perspectives of writing for publication in English. *The Asia-Pacific Education Researcher, 20*(3), 469–476.
Curry, M.J. (2003). Skills, access, and "basic writing": A community college case study from the United States. *Studies in the Education of Adults, 35*(1), 5–18.
Curry, M.J. (2011). "Where there is no laundry": Review of five books on writing for publication. *Journal of English for Academic Purposes, 20,* 1–4. doi:10.1016/j.jeap.2010.09.003
Curry, M.J. (2014). Graphics as invention heuristics in writing for publication by academic engineers. In M. J. Curry & D. Hanauer, (Eds.), *Language, literacy, and learning in STEM Education: Research methods and perspectives from applied linguistics* (pp. 87–106). Amsterdam: John Benjamins.
Curry, M.J. & Lillis, T. (2010). Making academic publishing practices visible: Designing research-based heuristics to support English-medium text production. In N. Harwood (Ed.), *Language teaching materials: Theory and practice* (pp. 322–345). Cambridge: Cambridge University Press.
Curry, M.J. & Lillis, T. (2013). *A scholar's guide to getting published in English: Critical choices and practical strategies.* Bristol, UK: Multilingual Matters.
Dicks, B., Flewitt, R., Lancaster, L. & Pahl, K. (2011). Multimodality and ethnography: Working at the intersection. *Qualitative Research, 11,* 227–237.
Gimenez, J. & Thondhlana, J. (2012). Collaborative writing in engineering: Perspectives from research and implications for undergraduate education. *European Journal of Engineering Education, 37*(5), 471–487.
Graves, H. (2005). *Rhetoric in(to) science: Style as invention in inquiry.* Cresskill, NJ: Hampton Press.
Hanauer, D. (2006). *Scientific discourse: Multimodality in the science classroom.* London: Continuum.
Hanauer, D. (2009). Science and the linguistic landscape: A genre analysis of representational wall space in a microbiology laboratory. In E. Shohamy & D. Gorter (Eds.), *Linguistic landscape: Expanding the scenery* (pp. 287–301). New York, NY: Routledge.
Harwood, N. (2005). What do we want EAP teaching materials for? *Journal of English for Academic Purposes, 4,* 149–161.
Hawhee, D. (2002). Kairotic encounters. In J. M. Atwill & J. M. Lauer (Eds.), *Perspectives on rhetorical invention* (pp. 16–35). Knoxville, TN: University of Tennessee Press.
Hutto, D. (2007). Graphics and invention in engineering writing. *Technical Communication, 54*(1), 88–98.
Kamler, B. & Thomson, P. (2008). The failure of dissertation advice books: Toward alternative pedagogies for doctoral writing. *Educational Researcher, 37*(8), 507–514.

Knorr-Cetina, K. (1981). *The manufacture of knowledge: An essay on the constructivist and contextual nature of science.* Oxford: Pergamon Press.

Koutsantoni, D. (2006). Rhetorical strategies in engineering research articles and research theses: Advanced academic literacy and relations of power. *Journal of English for Academic Purposes, 5*(1), 19–36.

Kress, G. (2011). 'Partnerships in research': Multimodality and ethnography. *Qualitative Research,* 11: 239–260.

LaFevre, K. (1987). *Invention as a social act.* Carbondale, IL: Southern Illinois University Press.

Latour, B. & Woolgar, S. (1979). *Laboratory life: The social construction of scientific facts.* Beverly Hills, CA: Sage.

Lauer, J. (2004). *Invention in rhetoric and composition.* West Lafayette, IN: Parlor Press.

Lea, M. & Street, B.V. (2006). Student writing in higher education: an academic literacies approach. *Studies in Higher Education, 23*(2): 157–72.

Leydens, J. & Olds, B. (2007). Publishing in scientific and engineering contexts: A course for graduate students. *IEEE Transactions on Professional Communication, 50*(1), 45–56.

Li, Y.Y. (2002). Writing for international publication: The perceptions of Chinese doctoral researchers. *Asian Journal of English Language Teaching, 12,* 179–193.

Li, Y.Y. (2005). Multidimensional enculturation: The case of an EFL Chinese doctoral student. *Journal of Asian Pacific Communications, 15*(1), 153–170.

Lillis, T. (2001). *Student writing: Access, regulation and desire.* London: Routledge.

Lillis, T. (2003). An 'academic literacies' approach to student writing in higher education: Drawing on Bakhtin to move from critique to design. *Language and Education, 17*(3), 192–207.

Lillis, T. (2008). Ethnography as method, methodology, and "deep theorizing": Closing the gap between text and context in academic writing research. *Written Communication, 25*(3), 353–388.

Lillis, T.M. & Curry, M.J. (2006). Professional academic writing by multilingual scholars: interactions with literacy brokers in the production of English-medium texts. *Written Communication, 23*(1), 3–35.

Lillis, T.M. & Curry, M.J. (2010). *Academic writing in a global context: The politics and practices of publishing in English.* London: Routledge.

Lillis, T. & Scott, M. (2007). Defining academic literacies research: issues of epistemology, ideology and strategy. *Journal of Applied Linguistics, 4*(1), 5–32.

Lloyd, P. (2000). Storytelling and the development of discourse in the engineering design process. *Design Studies, 21,* 357–373.

Luzon, M.J. (2005). Genre analysis in technical communication. *IEEE Explore,* 285–295.

Mathison, M. (2000). "I don't have to argue my design – the visual speaks for itself": A case study of mediated activity in an introductory mechanical engineering course. In S. Mitchell & R. Andrew (Eds.), *Learning to argue in higher education* (pp. 74–84). Portsmouth, NH: Boynton/Cook.

Nathan, M. & Petrosino, A. (2003). Expert blind spot among preservice teachers. *American Educational Research Journal, 40,* 905–928.

Odell, L. & Swersey, B. (2003). Reinventing *invention*: Writing across the curriculum without WAC. *Language & Learning across the Disciplines,* 6(3): 38–53.

Poe, M., Lerner, N., & Craig, J. (2010). *Learning to communicate in science and engineering: Case studies from MIT.* Cambridge, MA: The MIT Press.

Rymer, J. (1988). Scientific composing processes: How eminent scientists write journal articles. In D. A. Joliffe (Ed.), *Writing in academic disciplines* (pp. 211–250). Norwood, NJ: Ablex.

Selzer, J. (1983). The composing processes of an engineer. *College Composition and Communication, 34*(2), 178–187.

Shaw, P. (2010). Research reports in academic and industrial research. In M. F. Ruiz-Garrido, J.C. Palmer-Silveira & I. Fortanet-Gómez (Eds.), *English for professional and academic purposes* (pp. 73–87). Amsterdam: Rodopi.

Shehzad, W. (2006). How to end an introduction in a computer science article: A corpus-based approach. In E. Fitzpatrick (Ed.) *Language and computers, corpus linguistics beyond the word: Corpus linguistics from phrase to discourse* (pp. 227–241). Amsterdam: Rodopi.

Swales, J. (1990). *Genre analysis: English in academic and research settings.* Cambridge: Cambridge University Press.

Wenger, E. (1998). *Communities of practice: Learning, meaning, and identity.* Cambridge: Cambridge University Press.

Winsor, D. (1992). What counts as writing? An argument from engineers' practice. *Journal of Advanced Composition, 12*(2), 337–347.

Winsor, D. (1994). Invention and writing in technical work: Representing the object. *Written Communication, 11*(2), 227–250.

PART III

Research studies in mathematics

CHAPTER 7

Mathematics education, cultural practices, and communication*

Darlinda Moreira and Joana Latas
The Open University, Portugal / University of Évora

This chapter addresses the role of students' cultural practices in mathematical tasks as a means of enhancing discourse and communication in multicultural classes. It starts by presenting the features of mathematical discourse and highlighting their relationship with students' native languages and cultures. Next, the concept of cultural practices is discussed and framed in the context of mathematics education. Its potential to launch the construction of mathematical meanings is explored. The integration of students' cultural practices in the teaching and learning of mathematics is illustrated and discussed through the presentation of examples taken from classroom activities. These examples were collected from a research project that used a qualitative methodological approach to study the integration of students' cultural practices in the teaching and learning of mathematics. Seventh-grade students' cultural practices such as surfing were adapted and presented as activities that supported the learning of mathematical concepts. The chapter concludes by arguing for the contributions of students' cultural practices to the development of their ability to communicate mathematically, to use appropriate forms of language, and to connect mathematics to their daily experiences.

Introduction

The multicultural reality of today's schools poses new questions to mathematics educators, who must find ways to guide the learning of mathematics for all students regardless of their cultural, linguistic, and social backgrounds. Discourse,

* The authors are grateful to the editors for the invitation to collaborate on this book as well as for helping to edit the text. This help was invaluable and without it this collaboration would have been impossible.

communication, and culture act as important contexts for the understanding and development of mathematical knowledge and influence in various ways the mathematical performance of diverse students, including how they display mathematics' specific forms of knowledge. Indeed, since the 1970s, studies have shown the importance of cultural, social, and linguistic factors to the learning and teaching of mathematics. These studies have considered the sociocultural dimension of students' experiences and backgrounds to better understand and critically examine the complexity of pedagogical processes in multicultural classes. Research reveals that different cultural patterns of socialization and language systems lead to students' varying perceptions of mathematical concepts and relations as well as different types of engagements in the sociocultural practices that support school (Abreu 2005; Civil 2007; Moreira 1994; Secada, Fennema, & Adajian 1995).

In addition, research has pointed out that for students to become familiar with mathematics, they must participate in developing mathematical accounts, taking their own realities as the context and motivation for mathematical representation and problem solving (Abreu, Bishop, & Presmeg 2010; Gorgorió & Planas 2001; Secada et al. 1995; Williams, Black, Hernandez-Martinez, Davis, & Wake 2009). Students' involvement with learning also depends on their interests related both to what they know or have lived as well as to what they want to know and their perspectives on the future. In this sense, according to their social and political contexts, each student's interpretation of future opportunities – the cultural foreground – is central to the production of meaning and their involvement in learning. Research shows that students' cultural backgrounds play an important role in the learning process and that, consequently, pedagogical decisions that take these relationships into account will support the learning environment (Alrø, Skovsmose, & Valero 2009; Latas 2011; Skovsmose 2002).

Within this frame, in this chapter we present research on a pedagogical approach that takes account of both students' backgrounds and the cultural foreground to valorize students' cultural practices and identities in order to promote meaningful contexts that support mathematical learning. We assume that classroom mathematical communication is a process of cultural interactions among students and between students and teachers that plays a key role in the learning and consolidation of students' mathematical knowledge. From this perspective, students contribute their meanings in dialogue and, with the help of the teacher, deepen their mathematical knowledge by discovering how their daily experiences are connected to mathematics.

We start by analyzing the concept of cultural practices in the context of mathematics education. We then explore the potential of such practices to support the construction of mathematical meaning. Through the presentation of two mathematical situations, we next analyze and discuss the integration of students' cultural practices into the teaching and learning of mathematics and their impact in mathematics classrooms as a way to enhance learning through communication in classrooms. We conclude by proposing that sharing cultural experiences in the classroom, constructing a common language for the justification of the mathematical ideas, and establishing connections with mathematical topics all contribute to students' ability to communicate mathematically and consequently to learn mathematics.

Mathematical communication and students' cultural contexts

Communication requires both context and interlocutors. Indeed, underlying the communicative process there is a context and a social group that gives meaning to it (Gee 1990). In educational settings, the negotiation of meanings and the interactions established among students and between students and teachers during the communicative process highlight the social and cultural aspects of mathematical communication and the relationship between them. These aspects are especially important because of the cultural diversity in contemporary classrooms and the multiple multicultural values and experiences of students and teachers within them. Moreover, because in mathematical classrooms there are at least two linguistic contexts–one that arises from students' native languages (or varieties of language) and another that comes from the language of mathematics – the process of mathematical communication becomes highly complex as a result of the mix of different types of languages in both written and spoken forms. As Barton (2008, pp. 151–152) highlights, "In order for communication to happen, not only do we need relevant experiences to communicate about, but we also need to have a reason to communicate, and, just as important in this case, a need to communicate formally." Thus, not only are students' languages and contextual and cultural elements crucial for communication but the way that mathematical meaning is translated, articulated, and interpreted within written forms is also crucial to students' development of mathematical knowledge.

In terms of the integration of students' cultural backgrounds into learning mathematics, Boaler (1993) suggests a bidirectional influence between contexts and the processes of transferring students' learning from school mathematics to their out-of-school contexts and vice versa. Indeed, transfer processes are hindered or facilitated by factors such as word meanings, the artificial discourse often used in mathematical tasks, and how students' assign meaning to these contexts. Several examples show that not only do students often misunderstand the widespread use of specialist terms such as *quadrilateral*, *hypotenuse*, and *multiplicand*, but also that misunderstandings might happen when students encounter a "false friend," that is, a familiar word in students' native language that also belongs to specialist mathematics discourse, such as *root* or *operation* (Edmonds-Wathen 2012; Moreira 1994; Pimm 1987). In addition, students' lack of experience about the context presented in the articulation of a mathematical problem or the situation described in the problem can hinder their mathematical performance (Civil 2007; Dowling 1998; Moreira 1994).

It is beneficial to use students' familiar contexts in classroom tasks to facilitate the learning of mathematics, because it enables classroom social interactions to promote the development of connections between mathematical ideas and students' varying social realities. According to Bishop (2005), drawing on students' contexts helps emphasize the process and the outcome of shared meanings and consolidates the connections that students can make between their prior mathematical understandings and the reorganization of these ideas in relation to new knowledge. Thus, when a teacher provides the transition between different contexts (Abreu et al. 2010) using students' available languages, discussions can contribute to the development of learning, as they involve the exploration of mathematical ideas from different perspectives as well as the negotiation and construction of concepts from each student's prior meanings and knowledge (Bishop 2005; National Council of Teachers of Mathematics, 2007). Discussions in small and large groups that take place in an environment of mutual respect and trust provide room for critically and constructively defending positions and questioning others.

In addition, we need to emphasize the importance of teachers' decision making about the nature and variety of contexts available to use in classroom tasks and how relevant these tasks are to students' backgrounds (Alrø et al. 2009; Skovsmose 2002; Vithal & Skovsmose 1997). Selecting and exploiting a particular context for a mathematical task in order to create a successful learning environment is crucial to making the task relevant and significant for all students. In this regard, understanding the potential of a particular context to be explored in the classroom requires advance analysis of its possibility for learning, keeping in mind both the target group of students and the potential to establish connections between the context and the mathematical concepts and structures to be taught and learned.

However, an approach of using the local mathematical ideas and practices as curriculum tools assumes that teachers are capable not only of knowing about local practices but also of contextualizing them and discussing their potential uses and limitations. Indeed some of the basic assumptions useful for conceiving of relationships between mathematics and culture are (1) recognizing the classroom as a place with its own cultural perspectives and values; and (2) understanding students' cultural solutions to daily problems that are related to mathematics. Thus a productive pedagogical approach arises from the activation of the relationships between the outside world and the mathematics classroom. Indeed, making connections with students' prior knowledge, including about their daily lives, requires a pedagogical approach that encompasses students' interests, identities, and future expectations (Begg 2001; Gerdes 2007; Latas 2011; Moreira 1994, 2007).

Assuming that interactions and sharing of ideas are essential components in the construction of mathematical knowledge, it is expected that while communicating, students are engaged in the development of shared meanings. As Barton (2008, p. 125) argues, "Mathematics is not a world of triangles, symbols, rules of argument; it is a world of networks of people talking about ideas." The negotiation of meaning in the classroom involves the way mathematical knowledge is constructed as a result of sharing ideas, with the help of both students' native languages and formal disciplinary discourses, or registers of language. Thus, negotiation not only takes place at the level of each student's background, but also affects the foreground, namely, the individual interpretation of future opportunities to learn in the school context and to understand professional/occupational or other opportunities for the future.

In summary, to the extent that students experience exciting activities that give them the opportunity to speak, write, and read about their daily lives in connection to mathematics, students can learn to communicate mathematically and learn mathematics while communicating. In this environment, classroom mathematical communication becomes a context where students share and negotiate meanings among themselves and therefore develop and consolidate mathematical knowledge and practices that contribute to their mathematical learning.

Research methodology

To analyze students' cultural practices and their role in mathematics classrooms to foster the development of mathematical communication, in the rest of the chapter we describe and analyze mathematical examples taken from our ongoing research (Latas 2011; Latas & Moreira 2011, 2012), which has the goal of taking advantage of students' cultural contexts as possibilities for creating mathematical meaning in the classroom and therefore to improve mathematical learning.

The study focuses on exploring students' cultural experiences in the classroom from a mathematical point of view, as a way to make visible the mathematics implicit in their experiences. The project uses qualitative methodology and follows the interpretative paradigm (Stake 1995). In viewing 'reality' as a construct resulting from interactions among people, the reflexivity and adaptability of the qualitative approach enables a holistic perspective on each student's mathematical learning processes and a deep and systematic understanding of the nature of the cultural influences on this process.

Our research stance aligns with the interpretive research paradigm (Heath & Street 2008), which aims to understand 'reality' from the perspective of research participants. Our research represents this stance by identifying cultural mathematical practices through gathering information about the sociocultural environments of the participants and, throughout the design and implementation of a curricular project, to use students' sociocultural environment in the classroom to work out mathematical concepts. The project was developed in five phases. The main objective of the first phase was to observe local practices. The second phase consisted of analyzing with students the data collected in phase one. During this phase the teacher and researcher built five mathematical tasks that connected local practices to the seventh grade mathematical school content. The third phase consisted of the implementation in the classroom of the first four tasks. Phase four consisted of the formalization of the mathematical concepts that had been worked on implicitly and intensively during phase three. Finally, phase five consisted of a final task targeted at establishing a "bridge" to the first phase, that is, returning to cultural practices, but now enriched with a mathematical point of view, and thus extending students' cultural knowledge by connecting it to mathematics. In short, to implement the project we developed several mathematical situations to be used in classroom activities in order to explore students' cultural practices that could be related to school mathematics. The main goals of these activities were to guide students to make connections between school mathematics and their experiences in local society. Thus we collected data using participant observation methods in classrooms, interviews with students, teachers, and members of the local community, and the collection and analysis of school and community documents.

The project was developed in School A, where the second author teaches mathematics. School A is located in a seaside village in the south of Portugal, where the population triples in the summer and tourism accounts for up to 60% of economic activity. This region attracts large immigrant communities, which translates into significant cultural diversity in the community and schools. The school has an average of 460 students annually, of which 14% are students from an immigrant background, from 13 nationalities, with German and English as the most represented. This scenario highlights issues of a cultural nature of which the school is aware, including students' proficiency in Portuguese. According to class assessments and national mean grades, students in School A are underachieving in mathematics. In general, they belong to families with few economic resources and low school expectations and participation, although parents will go to the school for specific meetings upon request. Participants in the study included 18 seventh grade students in School A. Just over 50% of students' parents have a third-grade education. Less than 10% of parents had completed secondary school or pursued higher education.[1]

From informal conversations with students and teachers, we identified a set of themes as relevant in the local culture and economy: wind turbines, surfing, the sea, and fishing. These themes established the basis for the development of the activities carried out during the research project and that were offered to students during nine classes of 90 minutes each.

Students were placed in homogeneous groups based on their knowledge of the practices related to each theme in order to provide greater interaction and depth of knowledge in each small group. In addition, when the small groups made their presentation to the whole class, the diversity of each group tended to elicit participation by classmates who were unfamiliar with the practice that the other small groups discussed. In fact, because each group addressed different issues and jobs, other students actively questioned their peers who presented their work rather than being unsure of the relevance of their comments or fearing the reaction of the teacher (Alrø & Skovsmose 2002). Here we present episodes from the small group that completed activities on the theme of surfing.

Mathematical communication grounded in students' experiences as a way to learn mathematics

This episode focuses on surfing as an example of local culture in which students' experiences are grounded. Although not all students in the class surf, most students are familiar with the sport, as the school has a surfing club and some of the teachers surf. Students are also used to observing people who come to the village to surf. Thus surfing has become a common practice in the community and a subject of conversation between teachers and students.

In the classroom, references to the relationship between surfing and mathematics arose in connection to students' future job opportunities in the field of sports and consulting. In addition, students who are surfers regularly needed to analyze and interpret data from Internet sites that gave the direction of the wind. Figure 7.1 presents an extract (and its English translation) from a student group's

1. In 1986, schooling in Portugal became mandatory for children from six to 15 years old. In 2009 education became mandatory for children between six and 18 years old. Nonetheless, in 2004 only 60% of Portuguese youth between 18 and 24 years old have completed beyond ninth grade, although across the European Union, for the same age group, about 85% of youth have reached this educational level.

written report submitted at the end of the task implemented in the classroom. The task aimed to identify and describe strategies for determining the direction and orientation of the wind and students should write a report, summarizing their ideas, and highlighting the need to use mathematical knowledge in their surfing practice. Finally students presented it, orally, to the whole class.

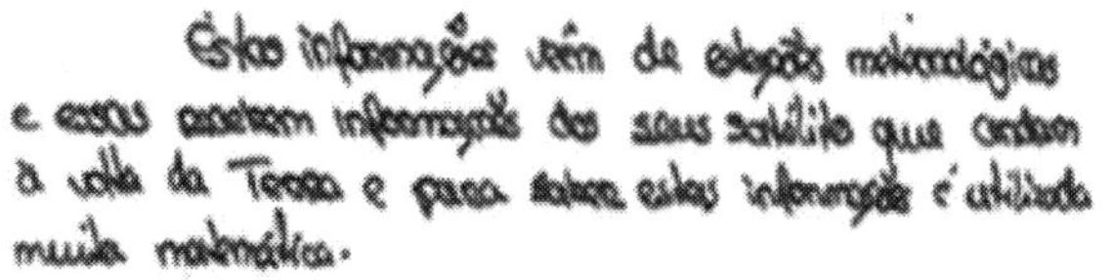

Translation: This information comes from meteorological stations and shows information collected from satellites that revolve around the earth and it uses a lot of math.

Figure 7.1. Extract of students' group work (and translation)

This example illustrates students' recognition of the role of mathematics in the world around them, although students recognized that they were not familiar with the mathematical knowledge used in creating the information identified about the applicability of mathematics their performance as surfers, that is, the mathematical knowledge acts at the level of their cultural foregrounds. Indeed students' comment that "[in meteorological stations] they use a lot of math" confirms their awareness that mathematics is present in relevant advanced technology, yet students failed to exemplify in practice the use of such mathematics, namely the knowledge of how to detect the direction of the wind, a skill needed in a sport like surfing.

In the next example, a task about daily cultural experiences was given to students in groups. The task presented situations related to the various local practices identified by students, and each group of students chose one situation. The surfing situation was chosen by the group of students who engaged in this practice.

The main goal of implementing this task was to present students with contextualized mathematical problems as an example of the use of formal mathematics in everyday practices. In addition, the overall objective of this task was for students to include an explanation of the mathematics involved in their cultural experiences and to establish connections between their lived experiences and classroom mathematics. Specifically, the objective was for students to identify folk measurements and expressions – in this case, the waves in the ocean – to relate them to the concept of estimation, then convert them into metric measurements and articulate mathematical concepts. The assignment given to students is provided in Example 7.1.

Example 7.1: Classroom activity based on surfing

> Tino is starting out in surfing. He wants to invest in a surfboard. He decides to buy a secondhand 8'5" board. Today Tino is going to try out the surfboard, but as a careful surfer he always evaluates the conditions before entering the sea. From the splashing of the waves he has determined that the wind is from off shore, that the swell is northwest, and that each set of waves is entering with a period of 12.
>
> Tino's analysis indicates that the waves are shoulder-high. However, his friend Roberto believes that the waves are one meter high, with some higher than a meter. Are these good conditions for surfing? Roberto, an experienced surfer, expects that it will be a good day because the waves are "tubular."
>
> From the available information, do you agree with Roberto? Justify your answer using words, diagrams, and symbols. Rewrite the above situation in a language that anyone could understand.

The problem in Example 7.1 aims to help students establish connections between their everyday world and mathematics. Cultural background was highlighted in terms of language used in the context of surfing, specifically, terminology that surfers would know, such as "tubular" to refer to the shape of a wave.

The task of rewriting of the problem by students triggered them to share their ideas in small group interactions as well as to develop respect for the knowledge of each group member. Conversations during this task gave rise to the use of various measures of length and their use in different contexts: Students translated the use of inches and feet for the length of the surfboard (following American mathematical practices) to the metric system in order for Portuguese students to understand this information. Figure 7.2 shows an extract of the problem text as rewritten by the group of students to fulfill the assignment.

The text in Figure 7.2 highlights that students recognize the importance of translating both the wind and ripple direction from the language used in the context of surfing to a common language understandable to all students. In addition, while communicating to perform the task, students were involved in the construction of mathematical meanings as well as the recognition of mathematical concepts, namely, estimation and regularity. In regard to estimation, students emphasized the role of the reference unit, pointing out that to the extent that the unit is more context related (authentic), it may be less accurate and therefore more subjective. In regard to regularity, students referred to the surfer's need to avoid wasting energy and therefore to predict the time to "catch the wave." Example 7.2 presents this debate in the classroom discourse.

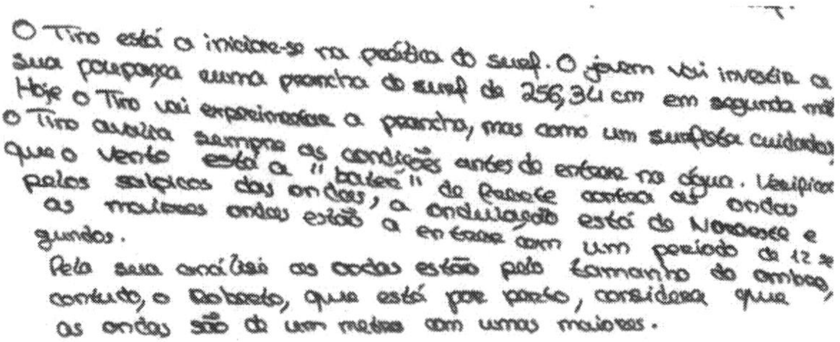
O Tino está a iniciar-se na prática do surf. O jovem vai investir a
sua poupança numa prancha de surf de 256,34 cm em segunda mão
Hoje o Tino vai experimentar a prancha, mas como um surfista cuidadoso
o Tino avalia sempre as condições antes de entrar na água. Verifica
que o vento está a "bater" de frente contra as ondas
pelos salpicos das ondas, a ondulação está de Nordeste e
as maiores ondas estão a entrar com um período de 12 segundos.
Pela sua análise as ondas estão pelo tamanho do ombro,
contudo, o Roberto, que está por perto, considera que
as ondas são de um metro com umas maiores.

Translation: Tino is starting out in surfing. He will invest his savings on a secondhand 256.34 cm surfboard. Today Tino will try out the surfboard, but as a careful surfer, before entering the sea, he always evaluates the conditions. He notes that the wind is "hitting" against the waves, splashing through the waves; the ripple is from the northeast and the largest waves are entering with a period of 12 seconds. According to Tino's analysis, the waves come up to his shoulder, however, his friend Roberto, believes that the waves are a meter high with some larger (Latas 2011, p. 102).

Figure 7.2. Extract of group A's rewriting (and translation)

Example 7.2: Classroom discussion of surfing and mathematical knowledge

Teacher: (to a small group of students): You said that you chose this situation to understand the relationship between mathematics and surfing and then ended up by saying that when Tony was surfing he used some mathematical knowledge, right?
Luisa: (to Tony): Do you use it?
Teacher: That was the question I wanted to ask. What do you think? Yes or no?
Students: Yes.
Teacher: Can you give concrete examples?
Luisa: The intensity of the wind?
Luis: The size of the waves.
Teacher: The size of the waves, how do you know?
Patricia: By estimating.
Teacher: By estimating. And in this case the estimate was made from what?
Tony: Shoulder height.
Teacher: Shoulder height. Therefore, the reference point was more or less the height of the shoulder.
António: But we can also say "head height," which is the height of the head and then we also have the waist.
Tony: We also use the period that is easier to estimate.
Teacher: What does "the period" have to do with math?
António: We count. We stay in the same place and we count until the next wave.

(From Latas 2011, pp. 102–103)

This dialogue brought out how different measures of length are used in different contexts, namely, the use of inches in reference to the length of the surfboard that was translated into the metric system. As Figure 7.2 shows, when students rewrote the problem texts to fulfill the assignment they did the conversion from 8'5" to 256.3 cm. The assignment also shows that students used the notions of direction and orientation to analyze the wind direction and characterized the ripple and the sea conditions in order to surf. When students rewrote the problem texts they use the equivalence between the expressions "offshore" and "the wind is" hitting "against the waves, splashing through the waves" or between "northwest swell is" and "the ripple is from the northeast."

The dialogue in Example 7.2 shows not only students' discussion in which they appropriate the mathematical notion of "period" but also how students use the notion of "regularity" to estimate the period between sets and to check if the size of the waves was acceptable for surfing. The presentation by the group to the whole class contributes both to the sharing of mathematical and cultural knowledge and to the enhancement of interactions between students with mutual respect for each other's knowledge. Concepts such as "shoulder height" or "waist height" gained meaning for students during the presentation by exploring the concept of estimation and discussing the average height of an adult. They drew meaning from the context during the negotiation of the mathematical topics involved in the situation. Thus, from a familiar context such as surfing students approached the mathematical notions of 'direction', 'orientation', 'estimation', and 'regularity' and were able to translate between different systems of measurements.

The discussion in Example 7.2 also highlights the relationship between students' cultural backgrounds and cultural foreground, as students use their cultural knowledge of surfing to engage in the use of mathematical tools to improve their surfing efficiency. In the classroom, students' curiosity was aroused about the relationship between surfing and mathematics, motivating dialogue among students who wanted to better understand the sport and the mathematics implicit in it. Moreover, interaction between students' cultural backgrounds and foregrounds promoted not only classroom debate but also students' investment in presenting their group's work to the whole class; therefore students' cultural foreground was crucial to fostering a repositioning of their interest toward mathematics. In fact, to the extent that their identification of relationships between the practice of surfing and mathematics developed, students constructed a new position from which to look at mathematics.

Discussion

In order to make mathematics accessible, grounded in students' experience and socioculturally embedded, students need to participate in constructing mathematical accounts, taking their own realities as the locus and leitmotif for mathematical representation (Moreira 2007). That is, the task of mathematics education should be to produce discourse with which teachers and researchers can interpret the complex daily realities of students and to transpose them into the language of mathematics. This task also demands the use of qualitative research not only to elicit from students the necessary information to ascertain their realities but also to better understand students' local practices to generate mathematical learning contexts and use them in classrooms. Indeed, exchanging viewpoints with students about how they perceive and experience the world around them helps teachers learn about their students and their relevant cultural practices that could be adapted and used for the teaching and learning of mathematics, rather than using artificial classroom activities of little interest to students.

To adapt cultural practices for students to work out relationships with mathematics in the classroom means that the teacher should simultaneously have a broad overview of the curriculum in terms of the articulation between mathematical topics. It also means that the teacher is involved with the mathematics curriculum in order to design innovative teaching materials. Thus, to argue for this stance for the teaching and learning of mathematics creates another issue, namely, the importance of the role of teachers as curriculum developers. In regard to Examples 7.1 and 7.2, students engaged in a cultural mathematical experience about surfing and accomplished the task of reformulating surfing language into daily language. Working with cultural aspects of mathematics related to aspects of surfing aroused students' curiosity and enthusiasm for the mathematics language and concepts identified in problem. Specifically, analysis of students' work on the task showed that they (1) recognized the presence of mathematics in their cultural knowledge and deepened that knowledge based on mathematical principles; (2) were involved in the process of making visible the mathematics entailed in cultural experiences and made connections with mathematical topics; (3) drew on their understanding of the physical world, such as the direction of wind and periods and height of waves, as well as related language, and (4) tried to relate situations from their everyday world of surfing using mathematical concepts.

Overall, using students' cultural practices contributes to the development of students' ability to communicate mathematically in the classroom. Students learn to employ appropriate forms of language to negotiate and share meanings that are understandable in both the context of their daily lives and in mathematical language, thus expanding the relationship between their cultural practices and their notions of mathematical applicability and of mathematics itself. In both examples, students' mathematical strategies became explicit and were useful for learning mathematics because of the possibility of mathematical sharing through the construction of a common language. For this purpose, students' exposure of their thinking and participation in class discussion, where they needed both to justify the ideas and appropriate meanings, clearly improved students' ability to communicate mathematically.

In conclusion, we note that teaching and learning mathematics must include what is intrinsic to students' lives: the culture of their local practices. Thus we argue for the inclusion of students' cultural experiences in school mathematics, by exploring, comparing, linking, and integrating these practices in order to construct content toward the enrichment of mathematics learning and intercultural education. We also argue for the importance of teachers to develop their knowledge of students' cultural practices and to be involved in developing authentic curriculum related to students' lives.

References

Abreu, G. (2005). Cultural identities in the multiethnic mathematical classroom. CERME-4 – *Proceedings of the Fourth Conference of the European Society for Research in Mathematics Education* (pp. 1131–1140). Sant Feliu de Guíxols, Spain, February 17–21.

Abreu, G., Bishop, A., & Presmeg, N. (2010). *Transitions between contexts of mathematical practices*. Dordrecht: Springer.

Alrø, H., Skovsmose, O., & Valero P. (2009). Inter-viewing foreground: Students' motives for learning in a multicultural setting. In M. César & K. Kumpulainen (Eds.), *Social interactions in multicultural settings* (pp. 13–37). Rotterdam: Sense Publishers.

Barton, B. (2008) *The language of mathematics: Telling mathematical tales*. New York, NY: Springer.

Begg, A. (2001). Ethnomathematics: Why, and what else? *ZDM – The International Journal on Mathematics Education, 33*(3), 71–74.

Bishop, A. (2005). *Aproximación sociocultural a la educación matemática*. (*Sociocultural approaches to mathematics education*) Santiago de Cali: Universidad del Valle.

Boaler, J. (1993). The role of contexts in the mathematics classroom: Do they make mathematics more "real"? *For the Learning of Mathematics, 13*(2), 12–17.

Civil, M. (2007). Building on community knowledge: An avenue to equity in mathematics education. In N. Nassir & P. Cobb (Eds.), *Improving access to mathematics: Diversity and equity in the classroom* (pp. 105–117). New York, NY: Teachers College Press.

Dowling, P. (1998). *The sociology of mathematics education: Mathematical myths/pedagogic texts*. London: The Falmer Press.

Edmonds-Wathen, C. (2012). False friends in the multilingual mathematics classroom. *Proceedings of the 11th International Congress of Mathematics Education* (pp. 5857–5866). Seoul, Korea, July 8–15.

Heath, S.B. & Street, B.V. (2008). *Ethnography: Approaches to language and literacy research*. New York, NY: Teachers College Press.

Gee. J. (1990). *Social linguistics and literacies: Ideology in discourses*. New York, NY: Falmer Press.

Gerdes, P. (2007). *Etnomatemática: Reflexões sobre a diversidade cultural*. (*Ethnomathematics: Reflections on cultural diversity*). Ribeirão, Portugal: Edições Húmus.

Gorgorió, N. & Planas, N. (2001). Teaching mathematics in multilingual classrooms. *Educational Studies of Mathematics, 47*(1), 7–33.

Latas, J. (2011). *O reconhecimento e a exploração da Matemática cultural: uma abordagem etnomatemática com alunos do 7.º ano de escolaridade.* (The recognition and exploration of cultural mathematics with seventh grade students.) Master's thesis. Lisbon: Association of Mathematics Teachers.

Latas, J. & Moreira, D. (2011). Uma abordagem etnomatemática em contexto de sala de aula. (An ethnomathematical approach in the classroom). *Proceedings of the 12th Interamerican conference on mathematics education* (XIII Conferência Interamericana de Educação Matemática). June, Universidade Federal de Pernambuco, Recife, Brasil.

Latas, J. & Moreira, D. (2012). Exploring connections between local and global mathematics. Topic Study Group 36. *Pre-proceedings of the 12th International Conference of Mathematics Education* (pp. 6982–6988). July, Seoul, Korea.

Moreira, D. (1994). *DJA: Mathematical conversations with a Portuguese speaking bilingual student.* Master's thesis. Lisbon: Association of Mathematics Teachers.

Moreira, D. (2007). Filling the gap between global and local mathematics. *Proceedings of the Fifth International Conference of the European Research Association on Mathematics Education,* pp. 1587–1596, February, 2007. Larnaca, Cyprus. ERME (European Society for Research in Mathematics Education) and Department of Education, University of Cyprus.

National Council of Teachers of Mathematics. (2007). *Principles and standards for school mathematics.* Reston, VA: National Council of Teachers of Mathematics. (*Princípios e normas para a matemática escolar.* Lisbon: Association of Mathematics Teachers).

Pimm, D. (1987). *Speaking mathematically: Communication in mathematics classrooms.* New York, NY: Routledge & Kegan Paul.

Secada W. G, Fennema, E. & Adajian, L.B. (1995). *New directions for equity in mathematics education.* Cambridge: Cambridge University Press.

Stake, R. (1995). *The art of case study research.* London: Sage.

Skovsmose, O. (2002). Students' foreground and the politics of learning obstacles. In M. de Monteiro (Ed.), *Proceedings of the 2nd International Congress on Ethnomathematics* (ICEM2). Ouro Preto, Brazil: Lyrium Comunicação.

Vithal, R. & Skovsmose, O. (1997). The end of innocence: A critique of ethnomathematics. *Educational Studies in Mathematics, 34*(2), 131–157.

Williams, J., Black, L., Hernandez-Martinez, P., Davis, P., & Wake, G. (2009). Repertoires of aspiration, narratives of identity, and cultural models of mathematics in practice. In M. Cesar & K. Kumpulainen (Eds.), *Social interactions in multicultural settings* (pp. 39–70). Rotterdam: Sense Publications.

CHAPTER 8

Learning while teaching

How classroom discourse practices mediate mathematics teachers' learning about student thinking

Jeffrey Choppin
University of Rochester

The literature in mathematics education has extensively documented the ways that discourse practices affect student learning and dispositions. However, there has been little discussion about how discourse practices affect what teachers learn in classrooms. This chapter builds from Davis's (1997) description of listening practices to explore the related propositions that what teachers learn in the context of their teaching is tied to the ways they listen, and their opportunities for listening are tied to the language patterns they enact in their classrooms. I provide examples of classroom discussions to illustrate various types of listening/language patterns and their implications for teacher learning. The episodes of classroom discourse demonstrate a continuum of teachers' listening practices and associated language patterns, from evaluative listening and teacher-centric language patterns to more interpretive listening and interactive discourse practices. In these episodes, evaluative listening involved less ambiguity than interpretive forms of listening, but revealed little about the deep ways students made sense of the mathematics. Interpretive and hermeneutic listening, by contrast, involved higher ambiguity but yielded deeper insights into student thinking than evaluative forms of listening. These findings highlight the importance of connecting teacher practices not only to student learning but also to what teachers know about how students learn mathematics. They also demonstrate the value of discourse analysis to understanding mathematics classroom practices of teaching and learning.

Introduction

The last two decades have seen increased attention to the social and cultural perspectives in research on mathematics teaching and learning (Lerman 2001).

These perspectives characterize learning by the ways people negotiate disciplinary knowledge and how they signify or communicate that knowledge (Latour 1988). Furthermore, the social contexts in which these practices are situated influence the goals and purposes of learning, and thus how people negotiate and signify their knowledge (Gee 1999). These perspectives challenge the transmission metaphor of learning, in part by describing knowledge as being contingent on the context in which it is developed and applied and by emphasizing the central role of language in learning. Sfard (2007, p. 565), for example, equates participation in discourse practices with learning, stating that "learning mathematics is tantamount to modifying and extending one's discourse." Substantial research has documented that learning opportunities for students are enabled or constrained by the language patterns in mathematics classrooms (see Herbel-Eisenmann, Choppin, Wagner, & Pimm [2012] for an account of this work). However, little attention has been paid to how language patterns affect what teachers themselves learn in mathematics classrooms, especially the knowledge they construct about how students learn in particular mathematical domains, which is crucial for adapting their teaching for specific contexts and students.

This chapter explores the related propositions that what teachers learn about students in the context of teaching is tied to the ways they listen to students, and that the opportunities teachers have for listening are tied to the language patterns in their classrooms. In typical mathematics classrooms, communication can often be characterized as one-way or monologic, with teachers attending primarily to the accuracy of students' responses to narrow questions that reveal little of students' thinking (Cazden 2001; Cohen 2011). Monologic language patterns provide little opportunity for the learner to influence the pre-determined path set out by the teacher or consequently for teachers to learn about students' perceptions of mathematical ideas. However, there is a growing literature that focuses on mathematics classrooms that feature more linguistically complex interactions (Goos 2004; Herbel-Eisenmann & Breyfogle 2005; Truxaw & DeFranco 2008). The goal of this chapter is to explore how these more complex language patterns may provide opportunities for teachers to increase their understanding of the ways students make sense of mathematics.

The chapter begins by describing a range of teachers' listening practices and connects them to classroom language patterns and ultimately to teacher learning about student knowledge or knowledge-making processes. It then provides episodes from classroom discussions to illustrate the various types of listening/language patterns identified in the framework. It concludes by discussing the implications of listening practices and language patterns for teacher learning and, ultimately, for student learning.

Teacher listening and classroom language patterns

Listening provides opportunities for teachers to understand how students make sense of academic content and how they learn it, and is thus critical to teaching that is responsive to learners (Schultz 2003). Through listening to the ways that students discuss mathematics, teachers can develop an understanding of what students think and how they come to think it, which allows them to connect students' informal and intuitive thinking to more conventional and disciplinary forms of reasoning (Ball 1993). Listening that is responsive to students' thinking is likely to be associated with more interactive language patterns than what is found in typical mathematics classrooms (Davis 1997). Notably, interactive language patterns provide opportunities for students to explain their thinking and respond to others' thinking, while in the process influencing the construction of knowledge and providing teachers the opportunity to develop an understanding of what students think in particular mathematical domains and how they come to think it.

Theoretical framework

Davis (1997) uses listening as a lens through which to describe various discursive patterns in mathematics classrooms. He describes a set of three broad teacher listening practices that represent a continuum in terms of the participatory roles they afford students. In the first, *evaluative listening*, the "primary reason for listening… tends to be rather limited and limiting" (p. 359), focusing primarily on the accuracy of students' answers. Furthermore, students' answers – right or wrong – typically have little bearing on "the prespecified trajectory of the lesson" (p. 360). Evaluative listening is strongly manifested in the initiate-response-evaluate (IRE) discourse pattern, which Cazden (2001) labels as the 'default' routine because it was so ubiquitous in the U.S. mathematics classrooms she observed. In the IRE pattern, the teacher typically poses a known-answer question, to which the student provides a brief response, which is then quickly evaluated for its accuracy (Cazden; Mehan 1979). In IRE-dominated exchanges, the teacher acts as the sole or primary arbiter of mathematical authority by providing the most important explanations or by leading students through a solution or explanation via a series of known-answer questions. As Lemke (1990) notes, the main purpose of this discourse structure is that "it gives the teacher almost total control of classroom dialogue and social interaction" (p. 168). In language patterns associated with evaluative listening, Davis notes, communication for the teacher is more about speaking than listening in that the teacher "is not really interested in what the [student] is saying" (p. 360).

The next two listening practices are linked to language patterns that are less structured and involve greater interaction with and between students. In *interpretive listening*, the teacher's attention to student thinking is "deliberate, and the [teacher] is aware of the fallibility of his or her sense-making" (Davis 1997, p. 364) in terms of ascertaining the content of students' thoughts and competencies. Interpretive listening moves the discourse away from the teacher as the sole authority and students having to guess the answer the teacher is seeking. In interpretive listening, the teacher asks questions whose answers cannot be fully anticipated, requiring an explanation or demonstration that involves sense making by students. Interpretive listening is based on the principle that "through interaction, we can affect one another's thinking," though, as Davis notes, "not necessarily in the manner intended" (p. 364). Interpretive listening is associated with more substantive student participation than evaluative listening because the teacher is focused on helping students to make sense of mathematical ideas. Davis explains, however, that because the teacher interprets student responses with respect to an anticipated set of possible responses, the teacher still largely dictates the content and flow of the discussion. Forman (1998), for example, described a teacher who typically facilitated a range of student solutions, in part due to her participation in a project that emphasized the use of open-ended problems to help students make sense of mathematics. In the episode analyzed for Forman's study, the teacher elicited a range of solutions but then focused only on specific features of the explanations. The data showed that the teacher attended only to a narrow range of explanations that matched her anticipated ways of approaching or discussing the mathematics.

The third practice involves the teacher paying greater attention to the "negotiated and participatory nature" of interactions in which multiple parties actively contribute to the collective development of ideas (Davis 1997, p. 369). In *hermeneutic listening*, the teacher elicits and organizes discussion around students' idiosyncratic forms of reasoning to develop their use of more formal and refined mathematical language, and more explicitly builds from the expansive variety of explanations offered by students. Compared with the previous two forms of listening, hermeneutic listening involves a greater degree of uncertainty in terms of the flow and trajectory of a class discussion, which requires that teachers develop responsiveness – what Davis calls a "particular response-ability" (p. 373) – to maintain a productive direction for the discussion.

The movement across the continuum represented in these three practices is toward more fluid, dynamic, and interdependent interactions in which ideas emerge as classroom participants collectively reflect on and discuss mathematics. Epistemologically, the evolution is away from conceptions of mathematics education "primarily as a system of [learning] already established, formal truths"

(p. 361) – whose goal is to avoid ambiguity – toward a complex, purposeful but unpredictable choreography that is geared toward reconciling students' and teachers' multiple and idiosyncratic ways of thinking about mathematics.

As language patterns progress toward more interactive forms, the norms related to classroom authority and social relations shift to foster more sustained and pedagogically productive discussions (Engle & Conant 2002). In addition, these norms can help participants hold themselves accountable to each other's ideas and to disciplinary forms of reasoning. Educational linguists studying language patterns in discussion-intensive mathematics classrooms propose the notion of *accountable talk* to describe linguistic features associated with the development and maintenance of these norms (Michaels, O'Connor, & Resnick 2008). The researchers identified a number of teacher and student moves that helped participants listen to and build from each other's ideas and to use disciplinary forms of evidence in their explanations. For example, teachers in their research persistently pressed students to explain both the technical and strategic aspects of their solutions to mathematical problems. Other moves provided opportunities for students collectively to negotiate how those solutions connected to underlying mathematical concepts. Teachers employed a move termed *revoicing* (O'Connor & Michaels 1993), in which they aligned or contrasted student contributions in ways that highlighted the underlying mathematical concepts. Typically, revoicing involved teachers (1) rephrasing or rebroadcasting a student explanation; (2) attributing the explanation's intellectual contributions to the student; and (3) checking with the student to see if the teacher had accurately described the explanation. Revoicing put the teacher "on relatively equal footing" with students (p. 324) and allowed students to "challenge or affirm" claims the teacher attributed to them. This shift in classroom power dynamics allowed the teacher to "induct students into a discourse community, by getting them to adopt roles in the ongoing thinking practices" (p. 325). Although teachers who are engaged in interpretive or hermeneutic listening may not intentionally use accountable talk moves, it is likely that these moves will be present in some form given how they function to elicit student thinking and to help teachers interpret and build from student explanations.

In sum, language patterns are associated with how teachers listen to students and influence what teachers learn about their students' learning while teaching. Typical language patterns in mathematics classrooms constrain opportunities for teachers to learn how their students make sense of mathematical ideas, with teachers listening for a narrow range of pre-determined responses. However, interpretive forms of listening entail more interdependent and collective negotiation of mathematical understanding, which can lead to more complex language patterns. Furthermore, the unpredictability of classroom discourse related to greater interactivity requires that participants develop social and disciplinary norms to maintain a productive focus on conventional mathematical ideas and terminology.

Research design

This study is situated within a larger research project that investigates teachers' understanding of curriculum materials and their instructional practices, especially discourse practices. In the larger research project, 12 middle school teachers were observed teaching instructional units from a set of innovative curriculum materials (Lappan, Fey, Fitzgerald, Friel, & Phillips 2006, discussed below). Data characterizing three of the 12 teachers' discourse practices are presented below.

Participants and their curriculum units

The episodes presented below occurred in the classrooms of three mathematics teachers working at the same middle school. All three were highly experienced in the use of the innovative curriculum materials and each had received extensive professional development with using the materials. Their school district used inquiry-based mathematics programs as primary curriculum materials for the entire K-12 grade span. Eighth grade teacher Amy Rudin[1] taught the *Say It with Symbols* unit, which focuses on developing understanding algebraic expressions. Cynthia Browne and Anabel Andrews were sixth grade teachers who frequently planned together for their accelerated sections, which used seventh grade units. Browne's episode involves the *Accentuate the Negative* unit, which focuses on computation with positive and negative rational numbers. Andrews's episode involves the *Comparing and Scaling* unit, which focuses on helping students develop methods for comparing quantities using multiple strategies, including fractions, ratios, and percentages.

Data collection

For the purposes of this research, I slightly modified the *teaching set* methodology (Cobb, Zhao, & Dean 2009; Simon & Tzur 1999) by creating *Unit Sets*. The Unit Sets consisted of six or more videotaped class sessions, which were usually consecutive. The class sessions were recorded from the beginning of the unit onward in order to capture how key concepts, representations, and procedures

1. The names of all of the teachers and students are pseudonyms.

were introduced and developed. The class sessions were parsed into segments according to the activity structure (e.g. homework review, lesson launch, group exploration, whole class summary discussion). The whole class summary discussions were transcribed and analyzed because they provided the most accessible opportunities to explore how teachers were listening and responding to students' strategies. The interactions between the teachers and students in the context of group work were often difficult to analyze because they sometimes involved questions and gestures about objects that were not easily captured by the camera. In addition to the videotaped class sessions, teachers were interviewed after each unit about their goals for the unit and perceptions about the implementations of the observed lessons.

Data analysis

The lesson transcripts of all 12 teachers in the larger study were analyzed to characterize interactional patterns, including the extent to which students provided explanations, and to document the presence of teacher discourse moves associated with accountable talk, including moves that prompted students to provide or evaluate a strategy, such as *probed students to explain reasoning*, *probed students to confirm or clarify an explanation*, and *elicited comments on a student strategy*. Other moves were analyzed to characterize how teachers responded to students' explanations, including *recalled past strategy*, *expanded on student response*, and *publicly displayed strategy*. Frequency counts were made across the 12 teachers' transcripts to characterize the ways teachers elicited and responded to students' explanations, yielding three rough categories of teacher practices. The 'low' accountable talk category included three teachers whose discourse practices included few of the moves listed above and whose students rarely provided explanations. The 'high' category included two teachers who, in comparison to the others, extensively used the accountable talk moves and whose students provided on average four times more explanations than students of the other teachers. The data presented below include one episode from a teacher in the low category and two from the teachers in the high category. The two teachers in the high category showed different tendencies in terms of how they responded to students' explanations, tendencies that mirrored the differences in the interpretive and hermeneutic listening practices. The data from each of these teachers was then searched to find episodes representing the broader patterns that could illustrate differences in the various listening practices described in the framework. The selected episodes were then analyzed to characterize in detailed ways how the teachers responded to student explanations in specific instances, which provided evidence for how the teachers were listening and responding to student thinking.

Curriculum materials

All of the episodes discussed below involve tasks from the Connected Mathematics Project (CMP) curriculum program (Lappan, Fey, Fitzgerald, Friel, & Phillips 2006). These materials emphasize student exploration in challenging tasks as a means of developing mathematical content knowledge. Each CMP unit comprises four or five *Investigations*, each of which consists of a task situation or context and several problems related to the same concept. Each Investigation follows a format of launch-explore-summarize. The launch familiarizes students with the context and the mathematics to prime student activity in the exploration, in which students engage with a task that is usually accessible through multiple approaches, which become the focus of the summary portion of the task. The summary discussion allows for students to compare various solutions and for the teacher to emphasize the important mathematical aspects of the tasks (Stein, Engle, Smith, & Hughes 2008).

Analysis of classroom episodes

In this section I present episodes from each of the three mathematics classrooms to illustrate evaluative, interpretive, and hermeneutic listening practices and the opportunities they provide for teachers to learn about students' thinking. The first episode illustrates the characteristic and affordances of evaluative listening; the second, interpretive listening; and the third, hermeneutic listening.

Evaluative listening

Episode 8.1, which takes place in Amy Rudin's classroom, illustrates the association between evaluative listening and the IRE language pattern. The episode involves the *Pool Tiling* problem in which students are asked to produce various expressions for a border of one-square-foot tiles around a square pool, with the ultimate goal of establishing the equivalence of these various expressions.

Episode 8.1: The Pool Tiling Problem

1. Rudin: Alright you guys got a whole bunch of expressions here that you say represents the number of tiles it takes to surround the pool. Now the question of course for A-3 [the problem number in the text] is what or how would I know whether these are equivalent to each other. How would I test these to find out whether they actually are equivalent? What did you do?
2. Annie: Substituted a value for *s*.

3. Rudin: Okay, substitute in a value for *s* and see how many tiles the expression calculates. Okay, what should you see if they are equivalent?
4. Annie: Get the same.
5. Rudin: Get the same what?
6. Annie: The same amount of tiles.
7. Rudin: Okay, so if the expressions tell the same number of tiles then they are?
8. Annie: Equivalent.
9. Rudin: Are equivalent or could be equivalent.

The exchanges above are typical of many mathematics classes in that the teacher asked a series of known-answer questions, sought a narrow range of responses, and corrected students when their answers did not fit the predetermined responses (Cazden 2001; Mehan 1979). For example, when Rudin asked, "So if the expressions tell the same number of tiles then they are?," she immediately affirmed the student's correct response of "equivalent." The episode continues:

10. Rudin: Okay, this is a big idea, guys, you want to make sure for A-3 that you have this written down, please. This is one way to go after this, this is one way to check to see if expressions are in fact equivalent. So somewhere on your lab sheet, somewhere on your grid sheets A-3 your answer for A-3 should look something like this. Alright, so then B focuses on the equations. It wants to turn these expressions into equations. Now, what do these equations calculate? Selena, what do these expressions give me?
11. Selena: Oh, the perimeter of the sides and how many tiles you're going to need.
12. Rudin: How many tiles I need.

In the above exchange, Rudin showed that she was "listening *for* something" as opposed to "listening *to* the speaker" (Davis 1997, p. 359, author's italics). For example, when she asked, "What do these expressions give me?," the student responded, "The perimeter of the sides and how many tiles you're going to need." This answer was partially correct – the border is not the perimeter – and Rudin simply repeated the correct part of the answer – "How many tiles I need" – in the next turn:

13. Rudin: Each of these expressions you are telling me tells me the number of tiles it takes to surround my pool. And what variable do they give you in *B* that represents the number of tiles it takes to surround the pool?
14. Selena: *n*.
15. Rudin: *n*. Okay, so what would I need to do then for B-1, what would I need to do to turn my expressions into the equations? Brian, how did you turn those expressions into equations?

16. Brian: You just do them.
17. Rudin: What do I do to my expressions to make them turn into equations, Jerry?
18. Jerry: Use the equal sign.
19. Rudin: Yeah.

The lack of responsiveness (what Davis termed "response-ability") in this set of interactions was evidenced in the exchanges above in the ways Rudin used the task sequencing in the book to structure the discussion (e.g. the explicit references to A-3 and B-1, which were labels of problems in the student text) rather than allowing students' solutions or explanations to influence the direction of the discussion.

The lack of exploration in the episode of Selena's misconception about perimeter demonstrates a missed opportunity to explore how that student and others were making sense of the task, which might have helped the teacher resolve students' confusion about the distinction between the number of border tiles and the perimeter of the pool. Similarly, the notion of equivalence was reduced to simply equating the number of tiles in each expression and not to the broader algebraic relationship between equivalent expressions.

From this interaction the teacher might have learned that a few students were able correctly to identify a procedure for determining equivalence and to recall certain terms. It would have been difficult, however, for her to learn how students were making sense of equivalence and algebraic expressions, which were ostensibly the two big ideas of the unit. For example, in turn 11 in the data, she simply asked the students to name the variable that the book provided to represent the number of tiles. Instead, she could have posed a question that might have provoked a more open discussion about the need to use a variable or what it means for expressions to vary with respect to a possible set of inputs. Consequently, the teacher was not responsive in ways that might have informed the trajectory of a more robust discussion or her own understanding of how students made sense of equivalence and algebraic expressions.

Interpretive listening

Episode 8.2 presents an example of interpretive listening in which the teacher attended to students' thinking. Although it involved greater degrees of uncertainty and student participation, it still had a strong controlling element on the part of the teacher. The episode took place in Cynthia Browne's class as she revisited a version of the *Temperature Change* problem, which required students to create number sentences to represent a situation in which the temperature changed from a starting value to an ending value. Browne directed the students to write number sentences that represented the problem situation, specifically in the form of *starting temperature + change = final temperature*, and not simply the computation used to find the change value. Before this episode students had established that subtracting a number is equivalent to adding its opposite, resulting in two equivalent number sentences. However, they had not established the equivalence between adding a number and subtracting its opposite.

Episode 8.2: The Temperature Change Problem

1. Browne: Okay what about in B? Negative 20 degrees to negative 10 degrees Fahrenheit. What would your number sentence be to represent that one? Natalie?
2. Natalie: Negative 20 degrees plus 10.
3. Browne: Correct. Do you have another number sentence for this one? Why don't we have another number sentence for this one? When I'm adding a positive is when I don't get that second number sentence.
4. Alan: But then you could subtract a negative.
5. Browne: I could subtract a negative, what do you mean?
6. Alan: That's exactly the same as adding a positive.
7. Browne: Okay did everyone hear what Alan said? Alan said, 'subtracting a negative would be the same as adding a positive.'
8. Alan: I think.
9. Browne: He thinks, now he's taking it back. Why do you say it is?
10. Alan: Well actually I don't have a good reason why.
11. Browne: He doesn't have a good reason why but he thinks it is.
12. Alan: No I know it is.
13. Browne: He knows it is. He doesn't even think it is anymore. Does anyone have a good reason why they might want to piggyback on what Alan just said? Kieran?
14. Kieran: Well if you subtract a negative number you're taking off, you're taking off two negatives just basically, well it's the opposite. If you add negative numbers you'll be (inaudible).
15. Browne: Okay wait. Kieran just said if you add a negative number it's like subtracting. So if you subtract a negative number it should be the opposite of that, so it should be like adding. That's a good thought. Matt?
16. Matt: Well if you're subtracting a negative amount then it's sort of like you're taking away something that was subtracted.
17. Browne: That's a good way of thinking of it. He said if you're subtracting a negative amount it's like you're taking away something that was already, I forgot what you said.
18. Matt: Subtracted.
19. Browne: Something that was already subtracted, so like you're taking it back almost. Okay and what were you just saying?
20. Alan: Oh I was just saying that I was sure that it is.

21. Browne: So now he's sure that it is. I like what Kieran said though, too, she said something about, can you repeat what you said Kieran? I forgot already, I really liked it though.
22. Kieran: They're opposites.
23. Browne: They're the opposite of each other. So if I am adding a negative it's like subtracting but because subtracting is the opposite of adding if I subtract a negative it should be like adding a positive.

This episode revolved around generating different interpretations for the claim made by Alan in Turn 4 that adding a positive is the same as subtracting negative. The two explanations made to support that claim (Kieran in turns 14 and 22 and Matt in turn 16) followed repeated efforts by Browne to press students to explain Alan's claim (turns 5, 9, and 13). Browne revoiced each explanation (turns 15, 17, 19, and 23), ending by emphasizing the use of 'opposite' to describe how the relationship between addition and subtraction mirrors the relationship between the sign of the numbers. This explanation is more formal than Matt's more intuitive "taking back" something that was subtracted, showing how Browne built from students' contributions to introduce conventional terminology.

Browne employed a variety of moves that generated more interaction than in the first IRE-dominated episode and which were responsive to student contributions, as evidenced in the way she acknowledged student turns 7, 13, and 21. She showed that she was attempting to make sense of student thinking, as evidenced in the revoicing moves mentioned above, and to connect that thinking to the major idea in which she was interested, the relationship between integer addition and subtraction. Browne stated in the post-unit interview that comparing the two number sentences helped students to make a critical connection between addition and subtraction and moved them beyond their experiences with whole numbers to understand and compute in new number systems.

The accountable talk moves that Browne employed, in particular the pressing moves, in which Browne probed students to explain their thinking, and revoicing moves helped to move students beyond merely making claims, as in Rudin's class, to providing evidence for claims. In fact, a key difference in this episode and others relative to IRE-dominated discussions was the relative lack of importance of students answering the initial question accurately. For instance, in turn 3, Browne simply replied, "Correct" to acknowledge that the student had generated a valid number sentence, then turned to the more substantial questions about why students had only produced one number sentence for this scenario.

Although this episode was more interactive than that in Rudin's class and relied on students' explanations to move the discussion forward, it still followed a closely determined path in that Browne was seeking a well-defined formulation of how adding a positive integer could be rewritten using subtraction. She stated in the post-unit interview that she had waited several days since a previous class when the students had established that subtracting a positive could be rewritten as adding a negative and she had chosen this moment to have students establish that adding a positive could be written as subtracting a negative.

From this interaction Browne could have gotten a sense of the ways that students were making sense of how negative numbers worked differently than positive numbers, which is a key difference between the number system students had previously worked with (whole numbers) and broader number systems with which students needed to develop familiarity, such as integers and rational numbers. Browne could also have learned more broadly about students' current and developing number sense, which is important for students as they move toward more advanced and abstract mathematics. Some of what Browne learned is likely tacit and informs her instruction in subtle ways, but the post-unit interview also revealed that she had developed a sense of how student thinking developed during a sequence of tasks related to integer addition and subtraction. From her observations of student thinking, which was made possible by the more complex language patterns that revealed how students made sense of the mathematics embedded in tasks, she subsequently made a number of important adaptations to the *Temperature Change* and other tasks in the sequence. These changes helped students to make important connections (reported in more detail in Choppin 2011a, 2011b).

Toward hermeneutic listening

Episode 8.3 took place in the classroom of Anabel Andrews and comes from an enactment of the *Apple Juice* problem. In the Apple Juice problem, students were asked to compare the 'appley' nature of four different mixtures of water and apple concentrate. The episode began with a student offering an explanation for how to find the strongest or weakest mixture.

Episode 8.3: The Apple Juice Mixture

1. Jeremy: You have to look at the proportions. Three to six isn't necessarily going to be equal to six to nine.
2. Andrews: And why not? So, yeah, that was the one with the difference. So, let's talk about B for a second. So, that's kind of what Jeremy has moved us to. If you look at how three to six, there's three cups difference … In Mix Y there's three cups difference, so would they be the same?
3. Students: No.
4. Andrews: No, so why not? Jeremy, what were you going to say?
5. Jeremy: In proportions, three to six is half, like if you were to make it a proper fraction thing, a proportion, uh three to six is half but six to nine is over half.

In this exchange, Andrews focused the students' attention on Jeremy's claim that "you have to look at the proportions" (turn 1) and then on his warrant for why the proportions were not equal, "three to six is half but six to nine is over half" (turn 5). The initial claim and warrant represent two key themes in the unit: first, comparing quantities proportionally rather than additively; and second, interpreting the quantities being compared. Jeremy's warrant involved a part-to-part comparison that was made in the language of a part-to-whole comparison. Andrews recognized this contribution as an opportunity to see how students were making sense of the quantities being compared and the nature of the comparison, which was evident in the ways she pressed students on the comparison statement that three cups concentrate is half of the six cups of water, as shown below.

6.	Andrews:	Okay, and so what's half? What do you mean, finish the sentence, half what?
7.	Jeremy:	Like, um, it's half of fifty percent. Half of fifty percent of the water.
8.	Andrews:	So, say it again. So, you got three cups concentrate to six cups water [writing *3 c. conc. to 6 c. water* on board]. Everyone listen to Jeremy because I'm going to ask you what he means by half when he says this.
9.	Jeremy:	Um, the three cups concentrate to six cups water, the three cups concentrate is about half of the six cups water. However, with the six to nine, six is over half of nine.
10.	Andrews:	Okay, so, okay good. Is half the mixture concentrate?
11.	Students:	Yes.
12.	Student:	No.
13.	Andrews:	Okay, talk in the group about what the half is.

After discussing in groups what the 'half' represented, the students offered two alternative interpretations of the quantity, first that the apple concentrate was one-third of the total mixture and, second, that for every six cups of cold water, there were three cups of apple concentrate. Andrews continued to press the students to relate their interpretations back to Jeremy's explanation, and a student ultimately explained that the quantity of the apple concentrate was half of the quantity of the cold water, which was the proper interpretation of Jeremy's 'half.'

As in Episode 8.2, more attention was paid to student reasoning than to the accuracy of answers. Andrews typically responded to students' answers with a fresh press for explanation or comment rather than evaluation. Not only did her persistent presses resulted in more frequent and more elaborate explanations than those in Rudin's and Browne's classes, but the explanations served as the primary objects of questioning and discussion.

Andrews' dependence on student explanations to drive the content and flow of the discussion made this episode more hermeneutic than the previous two episodes. Although Andrews had a clear idea of the ideas she wanted students to grapple with, she did not plan in advance to use this problem as the moment to press students. Instead, she responded to a particularly opportune explanation by a student that related to the key themes in the unit.

However, it was also evident that Andrews sought a particular interpretation of the quantities (concentrate to water) and of the nature of the quantities (part-to-part). Furthermore, she did not appeal to more intuitive or visual interpretations of the quantities, so it could be argued that the episode could have evolved in ways more true to a hermeneutic stance. The lack of even more responsiveness moves in part reflects the usual contingencies related to time and curriculum coverage and in part reflects a strategic choice on the part of Andrews to constrain the discussion while still being responsive to students' contributions.

From this episode Andrews could have learned that by pressing students to justify their responses, she could see how students were making sense of the key concepts emphasized in the unit. Her focus on Jeremy's explanation as an opportunity to press students to develop their explanations provided her with opportunities to learn about how students were thinking about proportions rather than differences, how they were attending to the quantities being compared, and how they understood the difference between part-to-part and part-to-whole comparisons.

Discussion

These three episodes demonstrate a continuum of listening practices and associated language patterns, from evaluative and direct to more interpretive and interactive forms. In each case, the teacher can be viewed as 'effective' depending on her instructional goals and in terms of what she may have wanted to learn about students. In the first episode, Rudin established a way of determining the equivalence of algebraic expressions; in the second episode, Browne established that adding a number could be rewritten as subtracting the opposite of that number; and in the third episode, Andrews established that the "half" referred to a part-to-part comparison involving cups of concentrate to cups of water. However, opportunities for students to explain their reasoning, and consequently for the teachers to understand how students were making sense of the mathematics, varied across the contexts.

Listening practices were implicated differently in the language patterns illustrated. In the case of Rudin, she clearly sought a narrow range of answers and lowered the ambiguity of interactions by asking students to complete sentences for which she provided the most critical parts. She was listening *for* particular answers instead of *to* the ways students reasoned, as was evident, for example, in the ways she ignored the confusion regarding border and perimeter. Browne, by contrast, elicited multiple student explanations, pressing and waiting until the terminology she sought was introduced, which provided her with an opportunity to revoice the students' explanations using a more conventional formulation. She demonstrated her responsiveness, and thus her interpretive listening stance, by attributing ideas to students in ways that maintained the substance and style of the students' claims. Andrews similarly pressed students to justify their reasoning, but in addition she allowed the students' explanations to dictate the direction and pacing of the discussion. She refrained from providing explanations, even in the form of revoicing, and instead relied on persistent presses to get students to provide the key.

The language patterns influenced how student thinking was revealed and consequently what teachers might have learned from the episodes. Evaluative listening may have helped Rudin to understand her students' command of procedures and facts, which is important but limiting in terms of designing instruction aimed at conceptual understanding. The evaluative language patterns yielded relatively unambiguous claims by students, while the interpretive patterns yielded more ambiguous claims that required additional work by the teacher to clarify. Interpretive and hermeneutic listening revealed more about student sense making, though it was more ambiguous (e.g. in the cases of Matt and Kieran, neither provided a mathematically complete and coherent explanation) and thus required higher degrees of inference. Andrews somewhat reduced the ambiguity by pressing students until they provided a clear explanation rather than revoicing and repairing students' explanations. In short, evaluative listening involved less ambiguity but revealed little about the deeper ways students made sense of the mathematics, while interpretive listening involved higher ambiguity but deeper insights into student thinking.

There is evidence beyond these episodes of the ways these teachers attended to student thinking and what they learned from it. Rudin's references to student thinking in the post-unit interview were mostly evaluative, with some descriptions of the details of students' solutions but no interpretations of what those details reveal in terms of student sense making. Browne and Andrews, by contrast, offered far more detailed and interpretive accounts of student thinking and linked these accounts to the ways students engaged with instructional sequences (see Choppin 2011a, 2011c for accounts of how teachers attended to student thinking). The ways that these teachers attended to student thinking, and what they learned from it, was evident in the ways they adapted tasks. Rudin primarily adapted tasks to make them less challenging by breaking them into smaller tasks and providing explicit instruction about how to complete the task, reducing ambiguity and thereby providing the opportunity for students to make key connections. Browne and Andrews, by contrast, adapted tasks in ways that addressed but did not remove the challenges and obstacles faced in previous enactments (see Choppin 2011a, 2011b, for a fuller account of these adaptations).

What teachers learn from their students while teaching has important implications for teachers' long-term growth. Ma (1999), for example, reports on how Chinese teachers develop 'knowledge packages' over time that comprise the teachers' interrelated understanding of mathematics and how children learn mathematics with respect to particular instructional sequences, a finding consistent with research on U.S. teachers who pay careful attention to student thinking (Franke, Carpenter, Levi, & Fennema 2001; Franke & Kazemi 2001). Such learning requires an epistemological orientation on the part of teachers that mathematics learning involves intensive, sustained, and supported student engagement in social practices (Davis 1997).

Implications for STEM educators

The examples in this chapter come from middle school mathematics classrooms, but they offer lessons that apply to STEM educators at other levels as well, the first of which is that interaction facilitates learning. This happens for multiple reasons. First, students understand the merit and accuracy of their ideas. Second, students engage with perspectives beyond their own, which serve as feedback and insight. Third, as this chapter shows, is through interaction that teachers understand how students make sense of ideas, which helps them craft instruction in more nuanced and effective ways. Listening, as shown above, is fundamental to developing interactive instruction. However, listening opportunities are contingent upon the extent to which instructors elicit student thinking, which leads to the next lesson, discussed below.

The second lesson for STEM educators is that there are some basic techniques for eliciting student thinking, but that attention needs to be paid to the purpose and focus of questioning. The technique most evident in this chapter is to press students to explain and clarify their reasoning. However, when done only as a general heuristic, asking students to provide evidence for their claims can become tedious. Furthermore, pressing for only technical details and facts leads to narrow forms of interaction. Instead, pressing was most effective in the examples above when it was geared toward big ideas, the concepts that drive a discipline. In the examples above, the big ideas included variable, number systems, and proportional reasoning. The more the teachers' questioning focused on these ideas, the more substantial the interactions became.

The final, and perhaps most vital, lesson is that to recruit more active engagement from students, educators need to be genuinely interested in what students think. Emphasizing interactions and eliciting student thinking is a pointless exercise if STEM educators desire only to obtain predetermined responses to their questions. The episodes above show how the teachers' intellectual engagement with student ideas was manifest in the linguistic features of the classroom discourse, and these features served as cues which spurred more active and substantial student participation.

References

Ball, D.L. (1993). With an eye on the mathematical horizon: Dilemmas of teaching elementary school mathematics. *The Elementary School Journal, 93*(4), 373–397.

Cazden, C.B. (2001). *Classroom discourse: The language of teaching and learning* (2nd ed.). Portsmouth, NH: Heinemann.

Choppin, J. (2011a). The impact of professional noticing on teachers' adaptations of challenging tasks. *Mathematical Thinking and Learning, 13*(3), 175–197.

Choppin, J. (2011b). Learned adaptations: Teachers' understanding and use of curriculum resources. *Journal of Mathematics Teacher Education, 14*, 331–353.

Choppin, J. (2011c). The role of local theories: Teacher knowledge and its impact on engaging students with challenging tasks. *Mathematics Education Research Journal, 23*(1), 5–25.

Cobb, P., Zhao, Q., & Dean, C. (2009). Conducting design experiments to support teachers' learning: A reflection from the field. *Journal of the Learning Sciences, 18*(2), 165–199.

Cohen, D.K. (2011). *Teaching and its predicaments*. Cambridge, MA: Harvard University Press.

Davis, B. (1997). Listening for differences: An evolving conception of mathematics teaching. *Journal for Research in Mathematics Education, 28*(3), 355–376.

Engle, R.A., & Conant, F. R. (2002). Guiding principles for fostering productive disciplinary engagement: Explaining an emergent argument in a community of learners classroom. *Cognition and Instruction, 20*(4), 399–483.

Forman, E.A., McCormick, D.E., & Donato, R. (1998). Learning what counts as a mathematical explanation. *Linguistics and Education, 9*(4), 313–339.

Franke, M., Carpenter, T., Levi, L., & Fennema, E. (2001). Capturing teachers' generative change: A follow-up study of professional development in mathematics. *American Educational Research Journal, 38*(3), 653–689.

Franke, M., & Kazemi, E. (2001). Teaching as learning within a community of practice: Characterizing generative growth. In T. Wood, B.S. Nelson, & J. Warfield (Eds.), *Beyond classical pedagogy: Teaching elementary school mathematics* (pp. 47–74). Mahwah, NJ: Lawrence Erlbaum Associates.

Gee, J.P. (1999). *An introduction to discourse analysis: Theory and method*. New York, NY: Routledge.

Goos, M. (2004). Learning mathematics in a classroom community of inquiry. *Journal for Research in Mathematics Education, 35*(4), 258–291.

Herbel-Eisenmann, B., Choppin, J., Wagner, D., & Pimm, D. (Eds.). (2012). *Equity in discourse for mathematics education: Theories, practices and policies*. New York, NY: Springer.

Herbel-Eisenmann, B.A., & Breyfogle, M.L. (2005). Questioning our patterns of questioning. *Mathematics Teaching in the Middle School, 10*(9), 484–489.

Lappan, G., Fey, J.T., Fitzgerald, W.M., Friel, S.N., & Phillips, E.D. (2006). *Connected mathematics 2*. Boston, MA: Prentice Hall.

Latour, B. (1988). Drawing things together. In M. Lynch & S. Woolgar (Eds.), *Representation in scientific practice* (pp. 19–68). Cambridge, MA: The MIT Press.

Lemke, J.L. (1990). *Talking science: Language, learning, and values*. Norwood, NJ: Ablex Publishing.

Lerman, S. (2001). Cultural, discursive psychology: A sociocultural approach to studying the teaching and learning of mathematics. *Educational Studies in Mathematics, 46*, 87–113.

Ma, L. (1999). *Knowing and teaching elementary mathematics*. Mahwah, NJ: Lawrence Erlbaum Associates.

Mehan, H. (1979). *Learning lessons: Social organization in the classroom*. Cambridge, MA: Harvard University Press.

Michaels, S., O'Connor, M.C., & Resnick, L.B. (2008). Deliberative discourse idealized and realized: Accountable talk in the classroom and in civic life. *Studies in Philosophy and Education, 27*(4), 283–297.

O'Connor, M. C, & Michaels, S. (1993). Aligning academic task and participation status through revoicing: Analysis of a classroom discourse strategy. *Anthropology and Education Quarterly, 24*(4), 318–355.

Schultz, K. (2003). *Listening: A framework for teaching across differences*. New York, NY: Teachers College Press.

Sfard, A. (2007). When the rules of discourse change, but nobody tells you: Making sense of mathematics learning from a commognitive standpoint. *Journal of the Learning Sciences, 16*(4), 565–613.

Simon, M., & Tzur, R. (1999). Explicating the teachers' perspective from the researchers' perspectives: Generating accounts of mathematics teachers' practice. *Journal for Research in Mathematics Education, 30*(3), 252–264.

Stein, M.K., Engle, R.A., Smith, M.S., & Hughes, E.K. (2008). Orchestrating productive mathematical discussions: Five practices for helping teachers move beyond show and tell. *Mathematical Thinking and Learning, 10*, 313–340.

Truxaw, M.P., & DeFranco, T.C. (2008). Mapping mathematics classroom discourse and its implications for models of teaching. *Journal for Research in Mathematics Education, 39*(5), 489–525.

CHAPTER 9

Improving girls' interest in mathematics through writing

Findings from a mixed-methods study

Edel M. Reilly
Indiana University of Pennsylvania

This chapter argues that one effective way to engage students, especially female students, in the learning of mathematics is to incorporate writing activities into mathematics classrooms. The chapter reports on findings from a mixed-methods research study which examined female and male middle school students' perceptions of using writing as a tool for learning mathematics. Two research questions were examined: (1) What are students' attitudes towards writing in mathematics classes? (2) What indications of mathematical knowledge growth and learner development can be found in student writing for mathematics classes? A survey instrument was administered to 293 middle school students. Using a Likert Scale format, the students responded to 28 survey items about their study of mathematics and their use of writing in mathematics classes. A t-test was used to analyze these quantitative survey results. Additionally, five open-ended questions on the survey asked what students had learned from using writing in their mathematics classes and how they viewed this teaching approach. These open-ended responses were qualitatively analyzed to provide a deeper understanding of students' views on the use of writing in mathematics classes. Ultimately this study offers evidence which suggests that writing can be a key to heightening female students' interest in mathematics.

Introduction

In 2010 the Center on Education Policy (CEP) released a study that claimed girls and boys are now performing equally well on state standardized assessments in mathematics (Chudowsky & Chudowsky 2010). These findings made national headlines and seemed to herald an end to the longstanding problem of gender inequality in students' learning of mathematics. While the CEP study holds promise of improvement, other research indicates that the problem of female underperformance in mathematics is far from solved (Beecher & Sweeney 2008; Hyde, Lindberg, Linn, Ellis, & Williams 2008; Neuville & Croizet 2007). Van de gaer, Pustjens and Van Damme (2008) report that female students have not performed as well as their male counterparts on standardized tests, findings contrary to the CEP findings. Especially important is evidence that shows that few women are pursuing careers in mathematics and science-related fields (Hill, Corbett & St. Rose 2010; Kurtz-Costes, Rowley, Harris-Britt, & Woods 2008). In fact, Xu (2008) claims that only approximately 15% or less of mathematics professors are female, with change coming slowly. In 2009, concern about the lack of women pursuing careers in science, technology, engineering, and mathematics (STEM) reached the United States House of Representatives. During a congressional hearing, Representative Daniel Lipinski, chair of the Congressional Subcommittee on Research and Science Education, argued:

> We are not producing enough scientists and engineers for the increasing number of technical jobs of the future. .... We must have women engineers, computer scientists, and physicists. By broadening the STEM pipeline to include more women and other under-represented groups, we can strengthen our workforce. (Lipinski 2009)

Unfortunately, female students continue to be underrepresented in advanced mathematics and science courses (Ceci, Williams, & Barnett 2009; Penner 2008). This chapter argues that one effective way to engage students, especially female students, in the learning of mathematics is to incorporate writing activities into mathematics classrooms. The chapter reports on findings from a mixed-methods research study which examined female and male students' perceptions of using writing as a tool for learning mathematics. The study draws on qualitative research methodologies used in the fields of literacy, composition, and applied linguistics. The study used a mixed methodology to provide both an inferential statistical analysis of students' reported perceptions and an inductive and interpretive analysis of students' writing. The study examined two primary research questions:

- What are students' attitudes towards writing in mathematics classes?
- What indications of mathematical knowledge growth and learner development can be found in student writing for mathematics classes?

Ultimately this study offers evidence which suggests that writing can be a key to heightening female students' interest in mathematics.

Writing as a tool for improving students' learning of mathematics

One of the biggest problems with the teaching of mathematics, according to Geist and King (2008), is the traditional pedagogy that so many mathematics teachers employ. According to Geist and King, this pedagogy consistently emphasizes "rote drill and memorization rather than active concept learning" (p. 43). They argue that what is needed to improve mathematics teaching is a "curriculum that is developmentally appropriate, individualized, and gender responsive" (p. 43). This challenge has persisted. According to Hennessey, Higley, and Chestnut (2012), "despite the importance of mathematics teaching, anecdotal and experimental evidence leads us to believe deep understanding and learning of mathematics may not be happening" (p. 187). This is partly because the teaching of mathematics remains focused on formula drill and practice. Hiebert, Morris, and Glass (2003) say the average classroom in the United States largely reveals the same methods of teaching mathematics today as in the past and so students continue to learn disappointingly little mathematics. The primary reason for this, according to these researchers, is that even though education programs encourage teachers to develop new pedagogical practices, teachers often abandon these practices when they face classroom problems.

Nevertheless, advances have been made in mathematics pedagogy. Among pedagogical innovations, the use of writing as a tool for teaching mathematics holds promise, but research on it remains limited. A small group of studies has provided evidence of the benefits of using writing to teach mathematics (Baxter, Woodward, & Olson 2005; Gibson & Thomas 2005; Ntenza 2006; O'Connell, Beamon, Beyea, Denvir, Dowdall, Friedland, & Ward 2005; Urquhart 2009). The main focus of these studies has been on describing the types of writing activities used in mathematics classrooms. Ntenza (2006) collected data from seventh grade mathematics classes in six schools in South Africa. Three schools were in rural settings and three were in urban settings. He interviewed a mathematics teacher at each school, examined mathematics textbooks and teachers' lesson plans, and analyzed students' written work for mathematics classes. Ntenza's research showed that students used writing in two major ways, "symbolic writing" and "mathematical writing" (p. 334). Symbolic writing involved working through routine mathematics exercises such as addition and multiplication. Mathematical writing was more linguistically complex, and so Ntenza placed these writing activities into 4 major categories: "direct use of language," (p. 340, "linguistic translation," (p. 334), "summarizing and interpreting" (p. 336), and "creative use of language" (p. 337). Ntenza's categories were based on earlier work by Davison and Pearce (1988) on junior high school mathematics classes that found that 74% of writing done by students in the survey consisted of merely "copying and transcribing information" from textbooks or from what teachers had written on the board. Ntenza's research found similar limited uses of writing in the six schools he studied. Like Davison and Pearce, Ntenza found students doing copying and transcribing of mathematical information, which he categorized as "direct use of language." He notes that one teacher required students to copy all of the teacher's writing into their exercise books. Only one of the six teachers in Ntenza's study "gave learners a few problems that would require translation from mathematical symbols into written language" (p. 355). Students were asked to keep journals in two classes, which required students "to summarize and interpret mathematics topics … learners were also required to state, amongst other things, their feelings and problems with regard to the mathematics topics and possibly give suggestions to their problems." Finally, Ntenza found that creative uses of language were limited to the urban schools, where students write investigative reports on a mathematics project. However, these assignments were "informal, optional and for enrichment activities" (p. 337) and so only a handful of students did them. Overall, very little substantive mathematical writing was taking place in the schools Ntenza studied.

More encouraging was research conducted by O'Connell et al. (2005) on efforts to develop students' mathematical reasoning and problem-solving skills through writing. During the course of one school year, the researchers worked with 22 teachers across five schools in Maryland, where teachers used journal writing. Initially, the teachers "gave little instruction or support to students in their writing endeavor" (p. 193). Over time, however, the teachers "began to provide students with support, such as graphic organizers, demonstrations and modeling, tips for improving written work, and specific lessons on ways to more effectively explain or describe ideas in writing" (p. 193). Students had to explain how they solved mathematics problems and to write about problem solving. Students were also asked to explain the thinking behind their answers and to explain mathematical concepts. The students' difficultly expressing their thinking in writing made the researchers and their teacher participants realize that students needed more direct instruction in "how to put thinking into writing" (p. 196). The teachers conclude that: "Incorporating writing into our mathematics lessons helped us stay focused on teaching for understanding. As we reviewed our students' work, we were able to see the level to which they understood each concept, something a correct answer had not always conveyed to us" (p. 198).

Brandenburg (2002) discusses the importance of writing as an instructional tool for advanced mathematics students. In weekly journals, Brandenburg had precalculus and calculus students summarize material studied in class, write letters to younger students explaining how to solve problems, rank topics covered in class from least to greatest importance and explain their rankings, and analyze problems in which the solution contains an error and then explain how the error was identified. By writing in these mathematics classes, "students learned to formulate and express their mathematical thinking in a clear-cut and substantive manner using correct vocabulary ... they learned to pinpoint confusion, compare and contrast mathematical methods, and ultimately deepen their understanding and retention" (Brandenburg, p. 68). While Brandenburg's research was not aimed at methodological rigor, it does illustrate some of the benefits of using writing as a pedagogy for teaching mathematics.

Gibson and Thomas (2005) studied the use of writing in a high school geometry class where students in were having "difficulty visualizing geometric reflections and transformations" (p. 108). To work on this problem, students were asked to design quilts using various triangular patterns. They then wrote a description of their quilt design for a quilting catalogue, making it both attractive sounding and explaining specific geometric terminology. According to Gibson and Thomas, students were highly engaged with this activity. The classroom teacher also reported that students asked many questions about manipulating the triangles in their quilt designs. On the final unit test, students were required "to identify the different types of transformations represented by given geometric figures" (p. 111). Although the researchers do not report specific assessment results, they do note that "a larger group of students identified these correctly on the test than had done so previously."

These studies show how writing can be used to help students gain stronger understanding of the mathematics they are studying, helping students work through concepts and helping teachers assess which ideas students have mastered and which are still challenging. In these studies, using writing helps students learn how to communicate their mathematical thinking.

Additional research has focused on the general benefits from using writing as a tool for teaching mathematics (Ediger 2006; Gibson & Thomas 2005; O'Connell et al. 2005; Pugalee 2004). Burns (2004) discusses how teachers and students can use writing as a formative assessment tool, providing a way for "students to reflect on their own learning and to explore, extend, and cement their ideas about the mathematics they study" (p. 30). These studies also suggest that mathematics teachers needed to be prepared to change or adapt their lesson plans if they see in students' writings indications of confusion about the mathematical concepts being introduced.

In one promising study, Baxter et al. (2005) looked at the effects of using writing when teaching mathematics to low-achieving students. Baxter et al. studied four students in a seventh-grade low-track math class. The teacher of this class required students to write in a journal on a weekly basis. Students were asked to respond in writing to open-ended prompts that initially focused on students' opinions and feelings, with topics like "Do you think that calculators should be used in math class?" (p. 121). Later, the prompts asked about students' mathematical thinking, such as: "Why is 0.3 greater than 0.003? Explain to a fifth grader how to place the decimal in a multiplication problem" (p. 121). After analyzing the journal writing, the authors "identified multiple instances where the students were able to explain their mathematical reasoning, revealing their conceptual understanding, ability to explain, and skill at representing a problem" (p. 119). Students noted in their journals that they liked being able to express confusion and difficulties with mathematical concepts in privacy rather than having to do so in a class discussion. Students also reported that the writing enabled them to have a one-on-one correspondence with their teacher and to discuss their feelings about the class and their performance.

The growth of research on writing and mathematics can be attributed in large part to the National Council of Teachers of Mathematics' (NCTM) publication in 2000 of *Principles and Standards for School Mathematics*. NCTM included in those standards a new communication standard on writing which said "Writing in mathematics can also help students consolidate their thinking because it requires students to reflect on their work and clarify their thoughts about the ideas developed in the lesson" (p. 61). This call by NCTM to integrate writing into mathematics classes has become regularly cited by researchers. It has also become a foundational principle for teachers working to reform mathematics curricula.

The positive effect of NCTM's standards is that writing has become somewhat more accepted as a pedagogical practice in mathematics classes. In fact, Seto and Meel (2006) claim that "one of the most significant changes in mathematics pedagogy over the past couple of decades has been the increasing use of writing as a pedagogical tool" (p. 209, as cited in Lopez 2010). However, more research on the use of writing to teach mathematics needs to be carried out if the practice is to be implemented effectively. A number of researchers have responded to this need and NCTM's call to emphasize the importance of writing and communication in the mathematics curriculum. Studies by McCormick (2010), Wilcox and Monroe (2011), Cooper (2012), and Doty (2012) have examined the use of writing as a pedagogy in mathematics classrooms. These scholars argue that when students use writing to learn mathematics, their learning is enhanced because it enables them to reflect on processes and to articulate them.

However, "despite the continued call for literacy instruction across the curriculum, secondary teachers have been reluctant to take up the cause" (Draper 2002, p. 520). Draper notes that efforts at reform have had mixed results because content and method of instruction are not always explained sufficiently, thus many teachers have been "left with a blurry view of how to create a math class that invites inquiry, initiates students' questioning, and provides students with the opportunity to explore and discuss as they construct their mathematical understanding" (p. 522).

Likewise, Cooper (2012) says that although writing can be an effective tool for teaching mathematics, "Unfortunately, math teachers are often preoccupied with students' ability to perform math procedures to the exclusion of math explanations, applications, and problem solving" (p. 80). Cooper cites studies by Millard, Oaks, and Sanders (2002) and Steen (2007) in support of this conflict between teachers' desires to teach procedures versus mathematical reasoning. Cooper advocates using technology, specifically chat, forums, and blogs, to facilitate students' writing about mathematics. She notes that: "Students can use these tools as mediums to communicate with one another, using reasoning and problem-solving skills to build representations of their knowledge and make connections between their prior knowledge and new knowledge" (p. 81).

Thus there is general agreement that more research on how to effectively integrate writing into the mathematics curriculum is needed. As already noted, this call for research is in keeping with the principles of the National Council of Teachers of Mathematics, which has argued that communication skills need to be made a central part of the teaching of mathematics. It is within this research gap that the present study is situated. It builds on earlier research in this area by examining middle level students' perceptions of writing in mathematics classes and considers those perceptions in light of students' genders.

A study of writing in middle-school mathematics classes

As noted, this study examined two primary research questions:

1. What are students' attitudes towards writing in mathematics classes?
2. What indications of mathematical knowledge growth and learner development can be found in student writing for mathematics classes?

Middle school students in grades six, seven, and eight were studied. This study reports on these students' perceptions of the benefits and drawbacks of using writing as a tool for teaching mathematics.

Mixed methods research methodology

A mixed-methods research approach (Gay & Airasian 2003; Hanson, Creswell, Plano-Clark, Petska, & Creswell 2005; Onwuegbuzie & Leech 2004; Ross & Onwuegbuzie 2012) was selected for this study to build on both quantitative research methods common to the discipline of mathematics and qualitative research methods common to the disciplines of literacy, composition, and language studies. This use of interdisciplinary research methods is also in keeping with the interdisciplinary spirit of this project that links the disciplines of mathematics and literacy.

Participating teachers

Potential teacher participants were drawn from attendees at a presentation on writing in mathematics classes by the researcher, as part of a Pennsylvania Council of Teachers of Mathematics (PCTM) conference. The researcher contacted teachers who attended this session by email, asking if the teachers would be interested in participating in this study. Three teachers agreed. To be eligible to participate in the study, teachers had to use writing in their mathematics classes on a weekly basis. The teachers also had to have been using this pedagogical approach for at least a year before the start of the study. Study participants were drawn exclusively from among Pennsylvania public school teachers, for two reasons. First, in Pennsylvania all students are required to write explanations of their mathematical reasoning as part of the annual state assessment of their mathematics achievement. This makes Pennsylvania students' ability to write about mathematics particularly important. Second, because the researcher had worked as a Pennsylvania public school mathematics teacher, this gave her background knowledge that could be of potential use in understanding study data obtained from the students of Pennsylvania mathematics teachers.

Participants' use of writing to teach mathematics

The teachers in this study believed that writing was a tool for improving students' learning of mathematics. This belief gave the researcher access to students who had been using writing as part of their natural mathematics learning environment. Participating teachers also were fully committed to, and experienced this teaching practice. One teacher had been using writing as a part of her mathematics teaching for two years; the second teacher for three years; and the third teacher for five years.

Obtaining naturalistic means that the data are more likely to reflect teachers' natural approaches to using writing to teach mathematics. However, teachers' writing prompts were not derived from the research literature on using writing to teach mathematics. Nonetheless, the teachers had done some reading on writing and mathematics and had good basic awareness of the benefits of this pedagogical approach as well as some of ways writing has been integrated into mathematics classrooms.

All participating teachers in the study required their students to write on a weekly basis in a variety of formats. Students were required to keep a notebook, which all three teachers referred to as a math writing portfolio. In this portfolio, students composed summaries, essays, and explanations of the mathematical concepts they were studying. Participating teachers reported that the main purpose of the summary and explanatory writing they assigned was to help students synthesize the material they were learning. Students were also asked to write reflective pieces. All of this writing was graded by the teachers with the exception of the reflective writing, which teachers wrote responses to but did not grade. The teachers commented that the main purpose of these reflective assignments was to enable students to verbalize their mathematics thinking.

Although each of the teachers had developed their writing activities individually, their approaches bear a strong resemblance to the pedagogy outlined by Huang and Normandia (2009). Huang and Normandia reported on data from a research project on how students learn to communicate about mathematics. While they note that students are generally good at describing actions, such as choosing a solution method or sequencing a set of solution steps, students have much more difficulty "expressing math content at the level of theoretical understanding" (p. 3), such as analyzing a problem or justifying a choice. Their study asked three research questions about student views and perceptions, one of which was, "How do students view the role of writing in relation to their understanding of the math content?" (p. 4). This question aligns directly with the research questions for the present study of students' perceptions of how writing affects their learning of mathematics.

Huang and Normandia studied one teacher's use of writing in teaching two mathematics classes of 25 students. The students attended a suburban private high school in New Jersey in a middle to high socioeconomic area. The researchers observed classes, interviewed students, and collected student work and course artifacts. The teacher in the study "involved students in writing mathematically from time to time" (p. 5) and assigned a formal take-home writing task. While students initially felt awkward writing about mathematics, they became more comfortable writing in their classes over time. Huang and Normandia claim that, "Analysis of student interviews situated in discourse analysis of classroom interactions strongly indicates that talking and writing is important to mathematics learning" (p. 12).

Huang and Normandia's study is the most direct model for the present study, which also looks at how students perceive the impact of writing on their learning of mathematics. This study looks at students who wrote regularly for an entire year, rather than for one short writing intervention. Students completed a variety of mathematics class writing activities. Because the writing activities were slightly different from teacher to teacher and varied over the course of a full school year, students were surveyed for their general perspectives on how this writing work affected their learning of mathematics. Rather than asking students how specific types of writing activities had affected their learning, they were given open-ended survey questions to discuss specific writing activities. The students also were asked to submit samples of writing they had done for their mathematics classes, which were analyzed using categories drawn from the work of multiple researchers in the area, which I discuss next.

Efforts to classify the types of writing that mathematics teachers typically use in their classes have been made by Urquhart (2009), Ntenza (2006), Lopez (2010), and Harbaugh, Pugalee, and Adams (n.d.). Urquhart identifies four basic purposes to writing in mathematics classes: to record learning, to solve mathematical problems, to explain mathematical ideas, and to describe learning processes. Ntenza categorizes mathematics students' writing as symbolic or mathematical. Symbolic writing is used for working with formulas and algorithms whereas mathematical writing involves transcriptions of source texts, translations of symbols into words, translations of others' words into personal explanations, and creative uses of language where students explore mathematical ideas. Lopez surveys a broad range of studies of writing and mathematics and concludes that these studies either discuss writing based on its purpose or the mode of writing done.

Harbaugh et al. analyzed 49 high school students' research reports in mathematics to develop a coding scheme. It classifies writing into these categories: lower-level knowledge structures (description, sequence); algorithmic, higher-level structures (classification, evaluation); procedural accounts; and experimental accounts.

The writing that teachers in this study had their students do to connects with the classification schemes described above. Synthesizing the classifications discussed above, the mathematical writing done by students in this study can be classified into four categories of purpose: descriptive, procedural, algorithmic, and creative. A full discussion and analysis of the writing done by students in the study is not offered here, but it seems useful to point out that the writing done for the present study aligns well with the writing done in similar studies in the past.

Research sites

This study was conducted in mathematics classes in grades six, seven, and eight in three Western Pennsylvania schools. Three mathematics teachers responded to my request to study their classes. The teachers came from school districts typical of the Western Pennsylvania region: rural, suburban, and urban. According to the United States Department of Commerce (2012), both the rural and suburban schools had primarily white student populations with less than 3% of students of black, Asian, or Hispanic/Latino ethnicity. The urban school, in contrast, had a 14% black student population, a 3% Hispanic/Latino student population, and less than a 1% Asian population. This made the urban school the most diverse school population in the study. Providing even greater contrast, however, was the general income levels of the school districts. Data from the Pennsylvania Partnership for Children (2013) shows that the suburban district was fairly affluent, with only 24% of students eligible for free or reduced school lunches. The rural school was much more economically challenged, with 51% of students eligible for free or reduced school lunches. The most economically challenged school district, however, was the urban school, which had 80% of students eligible for free or reduced school lunches. While these demographics provide some sense of the types of schools from which data were collected, this study did not focus on student performance based on ethnic background or income status.

The three teachers who participated in the study each taught five mathematics classes in which they used writing. This gave me access to 15 classes containing a total 386 students. Of these, 293 students (130 girls and 163 boys) chose to take part in this study.

Data collection: Student survey

After teachers indicated an interest in participating in the study, permission was sought from their superintendents to conduct the study at the teachers' schools. Next, students and their parents in these districts were contacted for permission for students to participate in the study. A cover letter describing the purpose of the study was distributed to the students in their mathematics classes by their teacher along with an informed consent form for parents and students to sign and return. This step was necessary since all the potential student subjects for the study were under the age of 18. Once permission for students to participate in the study was obtained, the students were given a survey instrument which was administered in the students' regular mathematics classroom environment. Students who chose not to participate in the study were given an alternative learning activity to work on while the study participants completed the survey.

A survey instrument designed to gather information on students' perceptions of the benefits and limitations of using writing as a tool for learning mathematics was the primary source of data. Using a Likert Scale format, students were asked to respond to 28 survey items focusing on their study of mathematics and their use of writing in mathematics classes. Quantitative data were collected from these items. In addition to the questions in Likert Scale format, the survey also contained five open-ended questions. These questions asked the students what they had learned from using writing in their mathematics classes and how they viewed this teaching approach. Students had an opportunity to describe what they believed was working and what was not working when about this pedagogical approach. Students were also asked to explain why a particular writing activity seemed useful. The final portion of the survey was used to collect basic demographic information from the student participants.

Student participants

In the 15 classes where data was collected there were 386 potential student participants for this study. Out of this group, 293 consented to complete surveys and all 293 did so, yielding a 100% response rate for those who consented to participate. Of the study participants, 66% were in eighth grade and 34% were in seventh grade. Table 9.1 provides an overview of the student participants. Slightly more of the survey respondents were male (56%) than female (44%). The majority of the students who took the survey were taking algebra classes, because 66% of the participants were in eighth grade and in the participating schools algebra was considered primarily an eighth grade class.

Table 9.1. Frequencies for grade level, mathematics class, and gender

Variable	**Categories** **Response number (valid percentages)**			
n = 293				
Class	*Math 7*	*Pre-Algebra*	*Algebra*	*Geometry*
	20 (6.8%)	103 (35.2%)	158 (53.9%)	12 (4.1%)
Grade	*Seventh*	*Eighth*		
	100 (34.1%)	193 (65.9%)		
Gender	*Male*	*Female*		
	163 (55.6%)	130 (44.4%)		

Data analysis

Students' responses to questions on the survey that were in Likert Scale format were analyzed using a t-test. Results of the t-test are reported first. Next, students' responses to questions that were open-ended were analyzed using both inductive analysis (Hatch 2002) and interpretive analysis (Hatch 2002; Thomas 2003). These results are reported second. The open-ended survey responses were analyzed to identify students' beliefs and attitudes about using writing as a tool for learning mathematics. Responses were compared by gender. These open ended responses were analyzed to confirm and expand upon results derived from the quantitative analysis of the responses students gave to the Likert Scale questions on the survey.

Students' attitudes toward writing in mathematics classes

Inferential statistics were used to answer the research question, "What are students' attitudes towards writing in mathematics classes?" A t-test was used to examine the effects of variables on male and female attitudes toward mathematics classes that used writing. The SPSS statistical analysis computer software package (Statistical Package for the Social Sciences, n.d.; Wellman 1998) was used for performing these statistical tests. An independent sample t-test of the responses was computed for each variable to determine the *p*-probability. Three variables were selected to examine students' enjoyment of writing in mathematics class, enjoyment of writing in other classes, and preferences for a mathematics class that used writing. Students' gender was also reported.

Based on this t-test, several conclusions can be made about how the gender of students impacts their attitudes toward writing in mathematics and other classes. The t-test analysis for each of the three variables resulted in rejection of the null hypothesis. The null hypotheses were that there would be no significant difference between genders for enjoyment of writing in mathematics class, for enjoyment of writing in other classes, and for preferences for a mathematics class that used writing. As Table 9.2 shows, the t-test analysis indicates a significant difference between genders for all three variables. That is, there is a significant difference between genders for students' enjoyment of writing in mathematics class; a significant difference between genders for students' enjoyment of writing in other classes; and a significant difference between genders for students' preference for a mathematics class in which writing activities are used.

Table 9.2. Inferential statistical summary for the variables of students' gender and attitudes toward writing in the mathematics class

Variable	Gender	N	M	SD	t
Enjoy writing in math class	Female	130	2.18	.80	.01*
	Male	163	1.94	.80	
Enjoy wring in other classes	Female	130	2.68	.96	.00**
	Male	163	2.01	.86	
Prefer a class with writing	Female	130	2.55	.75	.00**
	Male	163	2.28	.74	

$\alpha = .05$ *$p < .05$; **$p < .01$; ***$p < .001$

The mean data represented in Table 9.2 show that for each of the three variables, female students responded more positively to writing activities than did male students. This finding seems to indicate, therefore, that female students have a more positive attitude than males do toward writing in mathematics classes.

Recalling Cooper's claim that writing enhances student achievement in mathematics classes, the finding that females respond positively to using writing in mathematics classes is a very good sign. Mathematics teachers can use writing activities as a way to positively engage female students in learning mathematics. If female students' interest in mathematics is increased, this can help encourage more of them to continue to pursue mathematics study, where they have been underrepresented (Ceci et al. 2009).

This study did not attempt to prove that writing in mathematics classes actually did improve students' acquisition of mathematical knowledge. Instead, the study focused on students' perceptions of their learning because of the research discussed above indicating that female students are less likely than their male peers to pursue advanced mathematics study. If writing in mathematics classes increases student engagement in the learning of mathematics, as this study's data show, that is an important finding whether or not the writing actually enhances students' learning. If students are learning mathematics and also learning to enjoy the subject matter, then they are much more likely to pursue the subject in the future. If students are learning mathematics but losing interest in the subject matter, there is little hope they will pursue the subject long term.

Additionally, students answered survey questions about how they thought their knowledge of mathematics was increasing as a result of the use of writing in the course. This provides some indication that using writing in mathematics classes does enhance performance, an issue which is discussed in the next section.

Indications of mathematical knowledge growth

Inductive and interpretive analysis was used to answer the research question, "What indications of mathematical knowledge growth and learner development can be found in student writing for mathematics classes?" I analyzed students' responses to five open-ended survey questions, looking for common themes. Responses from the female and male students were tracked for comparison. Of the 293 students surveyed, 130 were female and 163 male. While all 293 students completed the Likert scale questions on the survey, not all students answered the five open-ended survey questions. Specific response percentages are noted in Table 9.3 and expanded on in the discussion that follows.

Table 9.3. Open-ended survey response rates by gender

Question	Number and percentage of female students answering (of 130 total)	Number and percentage of male students answering (of 163 total)
Impact of writing on test preparation	77 (40%)	39 (24%)
Change to learning of mathematics	42 (33%) Answered "no change": 5 (4%)	18 (11%) Answered "no change": 18 (11%)
Why writing is important	81 (63%)	44 (27%)

Female students' open-ended responses

The first open-ended question asked students to describe how a specific writing activity had changed the work they did for a mathematics quiz or test. Because the question was worded to elicit a descriptive answer, students either answered the question positively or left it blank. For this question, 40% of the female students surveyed wrote answers. Of the 77 female students who answered this question, all made comments indicating they felt the writing activities in their mathematics classes had helped them get better prepared for tests and quizzes. Explanations the female students gave included:

- By writing it helped me understand because I took it slowly, one step at a time.
- I could remember what I wrote and applied it to solving a test question.
- It is harder to explain things when I have to write it out, but since it is challenging I understand it more.
- Writing activities let you know how important it is to pay attention to questions and your answers. I didn't read a question carefully enough.
- Writing activities helped me practice in as much detail as I can for my test.

Students were also asked if they thought writing in mathematics class had changed the way the learned mathematics. Of the female respondents, 33% stated that they believed writing activities had helped them understand mathematics better. However, 4% of the female students said that writing had not had any effect on their learning of mathematics. For the 42 students who responded positively to this question, reasons listed for believing that writing improved their general learning of mathematics included:

- Writing helps you understand better 'cause a lot of the time it helps you relate to the real world.
- It helps because it explains some tougher parts of the chapter.
- It just makes you think about the things more. It gets in your head.
- You have to really think about each step and why you do each step.

The highest response rate for female students came in response to question: "What do you think is important about using writing to learn math?" For this question, 63% of the female students surveyed wrote answers, and all 81 students stated that writing was important for their learning of mathematics. Among the many reasons given by female students for why writing played an important role in their learning of mathematics were the following explanations:

- It is another way of learning it [mathematics]. Some people might be better at writing and learn math better.
- It also allows me to understand where I made my mistakes.
- If we didn't write then how can we learn? If we didn't write then how could we go back over our notes?
- To write down lessons and examples to help you when you get stuck on something.
- It goes further in depth.
- I guess it helps me to review what I've learned because most of the writing involves explaining how to solve something.

These qualitative extracts lend further support to the findings from the statistical data about the positive value for girls of using writing as a tool for learning mathematics.

Male students' open-ended responses

In general, male student response rates to the open-ended questions were consistently lower than the female response rates. When asked to describe how writing activities changed their work for a test or quiz, only 24% of the male students surveyed responded to the question. Nevertheless, the 39 male students who answered this question stated that writing helped them better prepare for tests and quizzes. These are typical responses from male students:

- I really didn't understand until I did the writing activity and then I did good on a quiz.
- It made me understand the material more and when we would have to use it.
- A writing activity helped me because I might have done something wrong and the writing helped me.
- It helped me understand the concept more.

The one survey question that generated a fairly good response from male students asked whether writing has changed the way the students learned mathematics. In answer, 46% of the male students stressed that writing helped them understand mathematics material better. An additional 11% of male students said that writing had made no change to how they learned mathematics. Most of the 75 male students who answered this question stated that writing gave them different options for solving mathematics problems. Like the females, the male students reported that writing allowed them to see practical applications of mathematics to the real world:

- It makes me understand when we will use math in real life.
- It has opened me to more options in doing things.
- It helps you learn different possibilities for a problem.
- It helped me look at a problem and read it carefully.
- Writing made math easier.

Male students also mentioned that writing required them to think more about the mathematics and the concepts they were studying. They also noted that writing forced them to practice the individual steps towards solving a problem. For male students, writing seemed to be a way of catching or preventing mistakes that might have occurred in their work otherwise. This view is reflected in the following comments by male students:

- I had to think a lot more than I would have if it was just strictly math.
- It helped me think things through better and I realized that I needed to think things through better.
- It made me think through the steps differently.
- I had to think in more detail.

Finally, only 27% of male students answered the question on what they thought was important about using writing to learn mathematics. All 44 of the male students who answered this question stated that writing did help them to understand mathematics concepts better. These male students also mentioned that writing allowed them to see the practical applications of math and that writing gave them another way of looking at math. Below are some comments that male students wrote:

- When you write you use more than just one or two sections of your brain, same with algebra, if you combine the two, you will learn more.
- It makes you see when you will really use math.
- We need a variety of ways to learn math. If it was the same all the time, it would be really boring.
- I think by reading your answer, you understand better. Sometimes plain numbers are hard to understand.

In general, a smaller percentage of the male students than female answered the open-ended survey questions. Overall, though, only 4% of the female students and 11% of the male students said they thought writing had not impacted how they learned mathematics. No students wrote negative comments about writing and mathematics. Instead, students either gave positive responses about the impact writing had on their mathematics learning or left the questions blank. Most important, 63% of female students discussed ways they thought writing was important to their learning of mathematics. This high percentage argues that female students do respond very favorably to the use of writing in mathematics classes.

Conclusion

With respect to gender and writing in the mathematics classroom, this study's survey findings show evidence that females responded more favorably than their male counterparts toward writing in mathematics classes. When it came to enjoying writing in mathematics class and preferring to take a mathematics class that used writing, females were again more favorable in their responses to these survey items than their male counterparts. Overall, when comparing the male responses to the female responses to the open-ended survey questions, much higher percentages of the female students viewed writing in their mathematics classes as valuable to their learning. Unfortunately, not every student, female or male, chose to answer all the open-ended survey questions. While this could mean that some negative views on the writing in mathematics class were not reported, based on the age of the students surveyed, it seems more likely that students merely chose not to answer the open-ended questions to save time. The open-ended survey responses clearly indicate that the female students felt more positively than their male counterparts about the extent to which writing improved their understanding of mathematics concepts.

Mathematics teachers have long been concerned with finding ways to improve the achievement of female students. Although some progress has been made on the gender achievement gap, more work needs to be done to encourage female students to pursue upper level mathematics classes and mathematics-related career paths. The findings of the present study suggest that female students had strong positive responses to using writing in their mathematics classes. Thus, one conclusion to be drawn from the present study is that efforts should be made to make writing a more widespread teaching practice in mathematics. Mathematics teachers need to be encouraged to see the value in using writing as a teaching tool and be taught how to use writing effectively in their pedagogy. They also need to be encouraged not to cut writing activities out of their classrooms in favor of more test preparation time. As the responses to the open-ended survey questions show, female students felt that writing in their mathematics classes helped them better prepare for tests and quizzes than when they did not write.

Thus both the quantitative findings of this study and the qualitative findings of this study together argue that students, particularly girls, find value in using writing as a tool for learning mathematics. The qualitative analysis of students' open-ended survey responses adds depth to the meaning that can be made from the t-tests. It is this mixture of research methods that helps show how valuable writing can be as a learning tool for mathematics students.

In order to link writing and mathematics, teachers must cross disciplinary boundaries. As Wilcox and Monroe (2011) note, research on writing and mathematics integration remains limited despite NCTM's call for efforts in this area. The present study tackles this disciplinary challenge by blending quantitative with qualitative research methods to enhance the linkage of disciplines for the overall enhancement of mathematics learning, especially for female students. When it comes to the problem of helping female students to succeed in mathematics classes, this chapter argues that writing may be one solution for which mathematics teachers have been searching. Or, to put it more simply: Writing + Mathematics = Learning.

References

Baxter, J.A., Woodward, J., & Olson, D. (2005). Writing in mathematics: An alternative form of communication for academically low-achieving students. *Learning Disabilities Research & Practice, 20*(2), 119–135.

Beecher, M., & Sweeney, S.M. (2008). Closing the achievement gap with curriculum enrichment and differentiation: One school's story. *Journal of Advanced Academics, 19,* 502–530.

Brandenburg, Sr. M. L. (2002). Advanced math? Write! *Educational Leadership, 60*(3), 67–68.

Burns, M. (2004). Writing in math. *Educational Leadership, 62*(2), 30–32.

Ceci, S.J., Williams, W.M., & Barnett, S.M. (2009). Women's underrepresentation in science: Sociocultural and biological considerations. *Psychological Bulletin, 135,* 218–261.

Chudowsky, N., & Chudowsky, V. (2010). *State test score trends through 2007–08, Part 5: Are there differences in achievement between boys and girls?* Retrieved from Center on Education Policy website: ⟨http://www.cep-dc.org/index.cfm?⟩

Cooper, A. (2012). Today's technologies enhance writing in mathematics. *The Clearinghouse: A Journal of Educational Strategies, Issues and Ideas, 85*(2), 80–85.

Davison, D.M., & Pearce, D.L. (1988). Teacher use of writing in the junior high mathematics classroom. *School Science and Mathematics, 88*(1), 6–15.

Doty, L.L. (2012). A mathematician learns the basics of writing instruction: An immersion experience with long-term benefits. *Primus, 22*(1), 14–29.

Draper, R.J. (2002). School mathematics reform, constructivism, and literacy: A case for literacy instruction in the reform-orientated math classroom. *Journal of Adolescent & Adult Literacy, 45*(6), 520–529.

Ediger, M. (2006). Writing in the mathematics curriculum. *Journal of Instructional Psychology, 33*(2), 120.

Gay, L. R, & Airasian, P. (2003). *Educational research: Competencies for analysis and applications* (7th ed.). Upper Saddle River, NJ: Merrill.

Geist, E.A., & King, M. (2008). Different, not better: Gender difference in mathematics learning and achievement. *Journal of Instructional Psychology, 35*(1), 43–52.

Gibson, M., & Thomas, T.G. (2005). Quilting blocks: Writing in the geometry classroom. *Mathematics Teacher, 99*(2), 108–111.

Hanson, W.B., Creswell, J.W., Plano-Clark, Y.L., Petska, K.S., & Creswell, D. (2005). Mixed methods research designs in counseling psychology. *Journal of Counseling Psychology, 52*(2), 224–235.

Harbaugh, A.P., Pugalee, D.K., & Adams, M. (n.d.). *A descriptive analysis of secondary mathematics students' formal report writing.* Retrieved from ⟨http://www.academia.edu/1352583/⟩

Hatch, J.A. (2002). *Doing qualitative research in education settings.* Albany, NY: State University of New York Press.

Hennessey, M.N., Higley, K., & Chesnut, S.R. (2012). Persuasive pedagogy: A new paradigm for mathematics education. *Educational Psychological Review, 24,* 187–204.

Hiebert, J., Morris, J., & Glass, B. (2003). Learning to learn to teach: An 'experiment' model for teaching and teacher preparation in mathematics. *Journal of Mathematics Teacher Education, 6,* 201–222.

Hill, C., Corbett, C., & St. Rose, A. (2010). *Why so few? Women in science, technology, engineering, and mathematics.* Retrieved January 16, 2014, from the American Association of University Women: ⟨http://www.aauw.org/resource/why-so-few-women-in-science-technology-engineering-and-mathematics-executive-summary/⟩

Huang, J. & Normandia, B. (2009). Students' perceptions on communicating mathematically: A case study of a secondary mathematics classroom. *The International Journal of Learning, 16*(5), 1–21.

Hyde, J.S., Lindberg, S.M., Linn, M.C., Ellis, A., & Williams, C. (2008). Gender similarities characterize math performance. *Science, 321,* 494–495.

Kurtz-Costes, B., Rowley, S.J., Harris-Britt, A., & Woods, T.A. (2008). Gender stereotypes about mathematics and science and self-perceptions of ability in late childhood and early adolescence. *Merrill-Palmer Quarterly, 54*(3), 386–409.

Lipinski, D. (2009). Lipinski examines ways to improve science, technology, engineering, and mathematics education participation among females. Retrieved from ⟨http://www.lipinski.house.gov/press-releases/lipinski-examines-ways-to-improve-science-technology-engineering-and-mathematics-education-participation-among-females/⟩

Lopez, S.C. (2010). *Towards evidence-based practices in mathematics instruction: Investigating the impact of writing on student ability to solve mathematics problems.* Unpublished doctoral dissertation. Chicago: Loyola University.

McCormick, K. (2010). Experiencing the power of learning mathematics through writing. *Issues in the undergraduate mathematics preparation of school teachers, 4,* 1–8. Retrieved from ⟨http://www.k-12prep.math.ttu.edu/journal/curriculum/mccormick01/article.pdf⟩

Millard, E. S., T. L. Oaks, and T. M. Sanders. (2002). *Improving student achievement through inclusion of problem solving in the math curriculum.* Arlington,VA: ERIC Document Reproduction Service No. ED 469 078.

National Council of Teachers of Mathematics. (2000). *Principles and standards for school mathematics.* Reston, VA: Author.

Neuville, E., & Croizet, J.C. (2007). Can salience of gender identity impair math performance among 7–8 years old girls? The moderating rile of task difficulty. *European Journal of Psychology of Education, XXII*(3), 307–316.

Ntenza, S. (2006). Investigating forms of children's writing in grade 7 mathematics classrooms. *Educational Studies in Mathematics, 61*(3), 321–345.

O'Connell, S.R., Beamon, C., Beyea, J.M., Denvir, S.S., Dowdall, L.A., Friedland, N.G., & Ward, J.D. (2005). Aiming for understanding: Lessons learned about writing in mathematics. *Teaching Children Mathematics, 12*(4), 192–199.

Onwuegbuzie, A.J., & Leech, N.L. (2004). Enhancing the interpretation of "significant" findings: The role of mixed methods research. *The Qualitative Report, 9*(4), 770–792.

Penner, A.M. (2008). Gender differences in extreme mathematical achievement: An international perspective on biological and social factors. *American Journal of Sociology, 114,* 138–170.

Pennsylvania Partnership for Children. (2013). *School lunch-students eligible for free or reduced-priced lunch.* Retrieved from ⟨http://datacenter.kidscount.org/data/tables/2720-school-lunch-students-eligible-for-free-or-reduced-priced-lunch⟩

Pugalee, D.K. (2004). A comparison of verbal and written descriptions of student solving processes. *Educational Studies in Mathematics, 55,* 27–47.

Ross, A., & Onwuegbuzie, A.J. (2012). Prevalence of mixed methods research in mathematics education. *The Mathematics Educator, 22*(1), 84–113.

Seto, M., & Meel, E. D. (2000). Writing in mathemaWtics: Making it work. *Primus: Problems, Resources, and Issues in Mathematics Undergraduate Studies, 16*(3), 204.

Statistical Package for the Social Sciences (SPSS). (n.d.). *Wikipedia, the free encyclopedia.* Retrieved July 5, 2013, from ⟨http://en.wikipedia.org/wiki/SPSS⟩

Steen, L. A. 2007. How mathematics counts. *Educational Leadership, 65*(3): 8–14.

Thomas, R.M. (2003). *Blending qualitative and quantitative research methods in theses and dissertations.* Thousand Oaks, CA: Corwin Press.

Urquhart, V. (2009). *Using writing in mathematics to deepen student learning.* Aurora, CO: Mid-continent Research for Education and Learning.

United States Department of Commerce. (2012). [City] *quick facts from US Census Bureau.* Retrieved from ⟨http://quickfacts.census.gov/qfd/states/42/4238288.html⟩

Van de gaer, E., Pustjens, H., & Van Damme, J. (2008). Mathematics participation and mathematics achievement across secondary school: The role of gender. *Sex Roles, 59,* 568–585.

Wellman, B. (1998). Doing it ourselves: The SPSS Manual as sociology's most influential recent book. In D. Clawson (Ed.), *Required reading: Sociology's most influential books* (pp. 71–78). Amherst, MA: University of Massachusetts Press.

Wilcox, B., & Monroe, E.E. (2011). Integrating writing and mathematics. *The Reading Teacher, 64*(7), 521–529.

Xu, Y.J. (2008). Gender disparity in STEM disciplines: A study of faculty attrition and turnover intentions. *Research in Higher Education, 49,* 607–624.

PART IV

Final words: Facilitating interdisciplinary collaboration

CHAPTER 10

Facilitating STEM education through interdisciplinarity

Funding, science, and applied linguistics

Cynthia Bauerle, Graham Hatfull and David Hanauer
Howard Hughes Medical Institute / University of Pittsburgh /
Indiana University of Pennsylvania

There is increasing pressure for STEM education to reform in the direction of student-centered learning approaches, active learning, interdisciplinarity, and data-driven evaluations and assessments of student learning. These requirements derive from the need for STEM education to facilitate economic development, national health, technological innovation, and appropriate demographic representation. These contingencies create the need and opportunity for the integration of applied linguistic expertise within STEM education. In this chapter, a senior project officer from a major funding institution, a leading scientist, and a seasoned applied linguist discuss and exemplify the role and conditions in which applied linguistics can facilitate different aspects of STEM education. The Phage Hunting Integrating Research and Education (PHIRE) program is presented as an example of a collaboration that constructed shared, interdisciplinary education projects designed to enhance the program's pedagogical mission.

The context for collaboration

Within the United States and across the world there is increasing understanding of, and urgency over, the question of how best to educate young people in the fields of science, technology, engineering, and mathematics (STEM) and how to diversify the populations that are currently involved in these fields. This sense of crisis results from the understanding of the importance and the role of STEM education in relation to economic viability, national health, technological innovation, and overall security of countries and the need for appropriate demographic representation from across society in these activities. This sense of urgency is found in a plethora of reports from major bodies, such as the *Bio 2010* report released by the National Research Council (2003, 2009); the American Association for the Advancement of Science's (AAAS) *Vision and Change* report (2010), published with the National Science Foundation; the Association of American Medical Colleges and Howard Hughes Medical Institute (AAMC-HHMI) joint committee report (2009), and the President's Council of Advisors on Science and Technology report (PCAST, 2012). All of these reports call for significant reform in STEM education in the direction of student-centered learning approaches, active learning, interdisciplinarity, and data-driven evaluations and assessments of student learning. An emerging consensus is calling for more "scientific" approaches to teaching, in which evidence-based practices inform the development of educational environments which encourage students to learn the nature and practice of science in authentic ways (Hanauer, Hatfull, & Jacobs-Sera 2009; Handelsman, Miller, & Pfund 2007). The importance of these pedagogies is based on the desire to create engaging and professional STEM experiences with the hope of increasing the number and retention rate of students who enter these fields. Funding agencies, universities, colleges, STEM departments, and individual faculty members are taking this challenge seriously and actively exploring how STEM education can be conducted in new ways.

Such changes within STEM education suggest that expertise from a range of disciplines needs to be used and integrated within a range of projects directed at leading the process of innovation in education. There is a history of interdisciplinary problem-solving in STEM where expertise is borrowed across fields to offer solutions. One way forward in addressing the challenge of improving and diversifying science education is the construction of interdisciplinary teams with the required expertise to allow quality innovative educational programs in STEM to be proposed, implemented, and assessed. This chapter is written by three writers who are well versed in the construction of such interdisciplinary teams designed to advance science education. Cynthia Bauerle is a molecular biologist and assistant director of Precollege and Undergraduate Science Education at the Howard Hughes Medical Institute (HHMI). She has overseen a series of projects designed to improve science education initiated and supported by HHMI. Graham Hatfull is the Eberly Family Professor of Biotechnology at the University of Pittsburgh and an HHMI Professor. He is the initiator of the Phage Hunting Integrating Research and Education (PHIRE) Program at the Pittsburgh Bacteriophage Institute at the University of Pittsburgh and the lead scientist for the Science Education Alliance–Phage Hunters Advancing Genomics and Evolutionary Science (SEA-PHAGES) program funded by HHMI and situated in more than 70 universities and colleges across the United States. David Hanauer is a Professor of English at Indiana University of Pennsylvania and an applied linguist who specializes in science education, literacy, and assessment. He is assessment coordinator and educational researcher for the PHIRE program and the assessment consultant for the National Experiment in Undergraduate Science Education (NEXUS) collaboration, an educational experiment funded by the HHMI. All three authors have worked closely together in a range of capacities related to science education reform.

This chapter is written in three voices. First Bauerle describes the HHMI Professors Program and the innovative features of the program that promote interdisciplinary approaches to science education among the community of HHMI professors. Then Hatfull discusses his understanding of constructing interdisciplinary teams in supporting science education initiatives, including specifically the incorporation of an applied linguist to address educational assessments. Finally, Hanauer addresses his understanding of applied linguistics and its potential role within STEM from his perspective as an applied linguist embedded in nationally recognized science education projects. The last section of the chapter, co-authored by all three writers, summarizes the aspects of our work that facilitated our collaboration.

Bauerle: Facilitating innovative science education through interdisciplinarity

HHMI is best known for supporting top biomedical investigators in the United States and internationally. In addition, through its science education programs, HHMI has provided grants to research universities and colleges that fund a variety of science education initiatives with a central goal: To educate and prepare future generations of leading scientists, science educators, and science literate citizens.

For more than two decades, HHMI has supported science education through competitive grants to institutions and individuals in support of initiatives focused on precollege, undergraduate, graduate, and medical education. This approach is modeled on the way HHMI has funded its scientific researchers, focusing on people rather than on narrowly determined research projects. Science education programming at HHMI is designed to support innovative approaches to preparing future scientists and enhancing science literacy. To date, more than $900 million has been awarded to strengthen the quality of science education through faculty-driven efforts at hundreds of colleges and universities across the United States. Our approach to science education parallels the institute's scientific mission to support "people, not projects," the belief that creative strategies can emerge when faculty have the freedom to pursue novel approaches to promoting scientific literacy or developing talented students to be future leaders in science and science education.

I came to HHMI in 2009 after almost two decades in academia to take on the role of managing the HHMI Professors program, a signature initiative launched in 2002 to empower accomplished research scientists significantly to expand their activities in undergraduate science education through initiatives that link research and teaching. Through my interactions with HHMI professors and ongoing evaluation of the impact of this program on the professors, their home institutions, and the national landscape for science education, I have come to appreciate the many innovative features of the program, which I believe have been integral to its success.

The essence of the HHMI Professors program lies in its recognition of the particular potential of individual scientists to pursue novel ideas and strategies to improve science education. The program is highly selective and identifies top academic scientists who are engaged in leading research in their fields. As scientific leaders, professors pursue leading-edge research, adopting novel approaches and crossing disciplinary boundaries. Professors are encouraged to employ this same question-driven impulse to pursue challenges in science education in novel ways. A good example of this is the work of Graham Hatfull, who has moved forward scientific understandings of mycobacteriophage, at the same time providing an innovative, cost-effective approach to initiating students into scientific inquiry (Hanauer, Jacobs-Sera, Pedulla, Cresawn, Hendrix, & Hatfull 2006).

Further, the structure of the program provides flexibility that enables professors to adapt and expand their programs over time. This flexibility encourages an investigative approach to educational reform in which professors try "experiments" from which they learn how to design effective programs. By selecting accomplished investigators, we leverage the status and reputation of top scientists to build national leadership for a broader impact on the implementation of science education across the United States and in other areas where the HHMI has influence. The full potential of this approach is reflected in the Society of HHMI Professors, a committed community of scientific leaders who pursue a variety of innovative approaches to address the most challenging issues in science education.

Many of the programs created by HHMI professors have become nationally visible models for improved practices. For example, the "scientific teaching" approach which advocates integrating the principles of scientific knowledge construction in relation to educational practices developed by Yale University professor Jo Handelsman and colleagues serves as a central concept in national conversations about undergraduate science education reform (Handelsman et al. 2007). UCLA professor Utpal Banerjee's "science deconstruction" method, which involves teaching students how to analyze and present research data, provides an effective model for classroom-based approaches to teaching the nature and process of science (Clark, Romero-Calderón, Olson, Jaworski, & Lopatto 2009). Professor Sarah Elgin's Genomics Education Partnership at Washington University in St. Louis supports a network of undergraduate student investigators engaged in authentic genomics research. The evolution of Graham Hatfull's PHIRE program into the nationally disseminated SEA-PHAGES initiative (described in the next section) is a compelling illustration of how ideas emerging from the Professors program have evolved to have significant impact on the science education landscape as evidenced by the presence of this program in more than 70 schools across the United States and the world. Indeed, the Society of HHMI Professors has emerged in the past decade as a group of national leaders speaking in one voice about pressing issues in science education reform (Anderson, Banerjee, Drennan, Elgin, et al. 2011).

While HHMI professors' objectives vary in terms of focus, objective, and approach, one shared characteristic of their programs is the construction of interdisciplinary teams to pursue project development and success. In addition, several aspects of the design of the Professors program support the development of inherently interdisciplinary approaches. First, the Society of HHMI Professors itself is scientifically diverse, comprising geneticists, plant biologists, microbiologists, biochemists, and bioengineers, among others. This context translates into multiple perspectives on science education; conversations among professors at annual gatherings of the society are creative and broad ranging. HHMI professors frequently pursue collaborative opportunities, working together on activities related to their own or shared projects. Perhaps most important, the freedom to innovate engages the group's most creative impulses, encouraging the professors to approach science education as a research project in itself, to pursue what they see as the most challenging issues and questions in the field, and to collaborate in order to assemble the expertise needed to build programs and advance understandings of science education.

Another element that may be identified as a driver of interdisciplinarity is the emphasis that HHMI places on evaluating the outcomes of the activities and projects it supports. Program evaluation in general, and perhaps specifically authentic assessments within educational contexts, are inherently interdisciplinary (Hanauer & Bauerle 2012; Hanauer et al. 2009). While HHMI professors bring expertise in scientific research and training to their programs, they may have had limited formal experience with educational theory and assessment at the outset of their award. Evaluating their programs means collaborating with specialists who contribute knowledge about how student learning may be demonstrated or about strategies that that yield the most useful information about impact of science education interventions. In this regard, the collaboration that has developed between Hatfull and Hanauer in the context of the PHIRE program is a fine example of how the HHMI professors program has encouraged interdisciplinary approaches to all aspects of project development, including evaluation and assessment.

The benefits of this approach are multiple: Professors have expanded their skills in pedagogy and assessment of learning and their program activities have contributed to advancing understanding through disciplinary-based education research. Such interdisciplinary working groups, embedded in academic science departments and dedicated to creating educational programs, provide some of the best examples of effective science education practice. To further explore how interdisciplinarity works in science education, one such project funded and supported by the HHMI Professorship program will be discussed in more detail in the next two sections by two of the central personnel in this collaboration – Graham Hatfull and David Hanauer. Initiatives such as Hatfull's PHIRE program serve as important national models of novel interdisciplinary team structures for launching educational innovation from inside the research-focused context of academic science departments.

Hatfull: Lessons from the PHIRE – Interdisciplinarity in an innovative science education program

Since I joined the University of Pittsburgh in 1988, a primary focus of my research endeavors has been the study of mycobacteriophages, viruses that infect hosts such as *Mycobacterium tuberculosis*, the causative agent of human tuberculosis (TB). Between 1988 and 2000 members of my laboratory and I determined the complete genomic sequences of four mycobacteriophages. As we began to learn that these sequences would prove informative about viral diversity and evolution, we realized that we needed to isolate many more mycobacteriophages from the environment to advance these studies. In the academic year 2000–2001 we engaged high school and undergraduate students in the discovery of new mycobacteriophages and coupled this work with genomic sequencing, an essential approach for defining the massive genetic diversity of viruses.

In 2002, I applied for and received funding from the HHMI Professors program to develop phage discovery and genomics as an introduction for novice students (interested but with no prior research experience or relevant expert knowledge) to authentic (i.e. publishable) scientific research. The motivations were two-fold: To give an impetus to the core research interest in defining viral diversity and to identify more effective ways of introducing students to the research environment and the research experience in general.

As the Phage Hunters Integrating Research and Education (PHIRE) program got off the ground, it seemed evident from personal experience that it offered a simple entry point for students to engage in research and that it was broadly accessible to students, regardless of their academic background. It also became clear that we were ill-equipped to understand to what extent the program is effective and why. The assessment goals of the newly initiated PHIRE program were not to understand student learning *per se*, but whether the opportunity provided them an effective research experience and, if so, what programmatic elements were important, how they could be honed, and what features might be transferrable to other projects.

At the start of the program, Marisa Pedulla, PHIRE program coordinator, met David Hanauer through a non-professional interest they shared. After hearing a talk from David, I was convinced that he could provide the critical expertise that we needed. The fact that David was an applied linguist who knew little microbiology and I was a microbiologist who had limited experience of applied linguistics was not a deterrent. We were interested in learning from each other. This perhaps is how the best interdisciplinary collaborations function: with open minds and a specific, actable, common interest.

For an initial period, David joined the laboratory observing what students and other researchers did, how they talked to each other, how they represented their efforts in notebooks and posters, generally absorbing the culture of a microbiology research laboratory. Through subsequent discussions and writing, David effectively articulated the learning process that students were engaged in. One element was his realization of the importance of the context and the powerful transition from accessible, concrete beginnings (the physical search in dirt for potential sites which have bacteriophage) to a highly abstract point (comparative genomic analysis) where students interpret the biological value of computer-based data (which consists of DNA genome sequences). This attribute became one of several identified key components of authentic research projects for novice scientists.

The common ground between David and me is more broadly grounded than the example described above. There is a plethora of questions we would like to address, such as how to measure student engagement, how to quantify changes in project ownership, how to represent the scientific concepts used in posters and presentations for research dissemination, how literacy is used to promote science education, and how scientific knowledge can be disseminated broadly. In this rich collaboration the garden of common ground seems to expand as the work progresses and new questions emerge for which a microbiologist and an applied linguist can seek answers.

This collaboration was initiated and advanced through the generous support of the HHMI. Although the financial component was important, it was not the only one. In 2002 when the program began, there were 20 inaugural HHMI professors across the United States. These people often shared little in terms of scientific interests, but all were committed to finding innovative interventions to advance science education. Over the next four years, the twice-yearly meetings were powerhouses of stimulating exchanges, and each of us came away enriched, excited, and more empowered to implement our programs. These group meetings had a positive effect on my collaboration with David, having primed my thoughts about the need to understand the learning process and the effectiveness the PHIRE program.

The frequent exchanges at meetings and by other means of communication with HHMI professors and program administrators provided an opportunity to articulate the core facets of the PHIRE program and the research opportunities it provides to novice scientists. The program is well suited for broader dissemination and in 2008 HHMI initiated the Science Education Alliance Phage Hunters Exploring Genomics and Evolutionary Science (SEA-PHAGES) program. After four years, about 3000 first-year students at more than 70 colleges and universities have experienced this program – and there are good assessment data showing that not only did the program do no harm to student learning, it effectively engaged them in the process of scientific discovery, a rare opportunity for first-year students. The early introduction into the program and lack of screening for academic 'gifts' enables a much-needed broadening of the diversity of students engaged in research, in terms of both academic and cultural diversity.

The SEA-PHAGES program offers a new landscape for the Hatfull-Hanauer collaboration, with an abundance of interesting questions, especially those focusing on implementation of programs such as PHIRE and PHAGES in a wide range of academic contexts. Together with the continued engagement of HHMI, the intersection of microbiology and applied linguistics promises a rich future for addressing the core challenges in science education.

Hanauer: Applied linguistics and science education

For the last 10 years, I have worked as an applied linguist in a variety of science education research projects and initiatives mainly within the life sciences. These include the PHIRE program initiated by Hatfull (discussed above) and the National Experiment in Undergraduate Science Education (NEXUS) collaboration which involves new approaches to competency based undergraduate science teaching in biology, chemistry, medicine, and physics (a project directed by Bauerle). Broadly, my role has been to develop scientists' expertise in education and assessment and to provide data on the student learning outcomes of the educational projects I have been involved with. In other words, the vast majority of my research and work comes under the heading of science education and assessment and it is in this context that my skills as an applied linguist have proven useful.

Before I address the interaction of applied linguistics and science education, it is important to specify my understanding of what applied linguistics is and how it works. Applied linguistics covers a wide range of activities from language teaching to exploring the role of language in authentic settings. However, across its manifestations applied linguistics is characterized by its interdisciplinarity and commitment to providing practical solutions to real world, language-related problems. The interdisciplinarity of applied linguistics consists of the willingness of, and need for, applied linguists to borrow and incorporate theories and methodologies from a variety of related fields of inquiry. Applied linguists, as might be expected, employ linguistic theories of various types. But in addition they may incorporate theories and methodologies from psychology, sociology, education, literature, cultural studies, history, and computer science. The interdisciplinarity of applied linguistics is not limited to its sources, but also includes its applications. Applied linguists are involved in studying a diverse group of people from scientists to poets to lawyers to politicians as well as non-professionals. In this sense, it is not unusual for an applied linguist to be involved in an interdisciplinary enterprise such as the one in STEM. However, what is unusual in my situation is the degree of my integration and participation within the Hatfull laboratory for the last seven years. The collaboration with Hatfull and the continued presence with the laboratory setting have allowed a degree of embedded interaction that offers me an unusual perspective.

An equally important additional feature of applied linguistics is based on the fact that a vast number of human activities require linguistic interaction. Science, for example, requires the documentation of experiments, the construction and representation of findings, and the communication of results across a range of different modes and mediums such as laboratory notebooks, conference posters, invited lectures, and laboratory notebook meetings. In other words, science requires language and literacy and, as with other areas of linguistic interaction, a number of communicative problems may arise. It is these communicative issues that applied linguistics addresses. Questions such as how children understand marine life biodiversity or what undergraduate students know about the scientific process are issues of science education and representation. Identifying what a student knows and how we know what a student knows requires careful control over the available resources of communication. More broadly, applied linguistics can help with any situation that requires communication: for example, how scientists talk and explain their findings to students, politicians, newspaper reporters, and the public. In this sense, applied linguistics involves addressing communication and representation in real-world situations, exploring how these situations function, and offering solutions to enhance communication.

In my work as the assessment coordinator and educational researcher in Hatfull's PHIRE program, my background in applied linguistics directed my co-researchers and myself to a series of research questions about the knowledge developed by students who participated in this authentic educational experience. What are students learning? What conceptual changes take place in students during the lengthy process of participating in an authentic inquiry process? How do they use laboratory notebooks? How does the social setting and the type of laboratory framework contribute to the understanding and learning of science? What are the affective aspects of science education? How does the use of laboratory space enhance undergraduate science education? In what ways might students engage with and feel ownership of their scientific activities? What identities emerge while working in a professional laboratory setting? These questions and others were formulated through close discussion, interaction, and collaboration with Hatfull and educational coordinator Deborah Jacobs-Sera and led to the development of innovative approaches to the assessment and understanding of student learning within educational programs of this type.

Specifically, we developed an assessment approach termed *Active Assessment,* which incorporates principles from performance testing, the importance of context, diversified knowledge sources, the centrality of scientific inquiry, and the educational training of scientists (Hanauer et al. 2009). Interestingly, this approach exemplifies the principles of interdisciplinarity that underpin its conception and application. At the heart of this approach is ethnographic observation of scientific practices, deep appreciation of the science at the heart of the activity, and explicit linguistic and multimodal description of the processes involved. Together this descriptive process is translated into a number of instruments that can be used to assess and evaluate how and when learning takes place. We have published the results of these studies to make the educational knowledge we have generated accessible and useful to a wide audience of scientist-educators (Hanauer et al. 2006; Hanauer et al. 2009).

My work in the field of science education has been facilitated by the fact that applied linguistics offers an informed understanding of the modes and mediums of communication, combined with a broad range of theories and methodologies suitable for investigating student understanding and knowledge construction. As an applied linguist, I have had access to a set of tools and theories that can be directly applied to questions of educational design and assessment in STEM. As exemplified in my work and in the chapters of this book, applied linguistics – with its inherent interdisciplinarity, multifaceted approach to research, and focus on how education takes place – can provide a valuable resource to those working in STEM education.

Facilitating STEM education through interdisciplinarity

As might be clear from the perspectives presented above, the authors of this chapter believe in the value of interdisciplinarity as a way of innovating science education. We have evidence for its value in the form of shared collaborations, innovative educational programs, and research publications. As illustrated in our collaborations, applied linguistics as a discipline can provide a fruitful conceptual and methodological platform for the exploration of student learning. The willingness of all parties – from funder to lead scientist to applied linguist – to listen, learn, use, and creatively interact with shared questions of interest has helped facilitate this collaboration and led to constructive outcomes in the form of program design and research publications. However, the writers of this chapter are also aware that it is uncommon for an applied linguist to be directly involved with STEM education; thus we consider it worthwhile to finish by summarizing the aspects of our work that facilitated our collaborations.

1. *Recognizing the need for change in STEM education*: As a basic underpinning issue, our collaboration was facilitated by the recognition that new ways of conducting STEM education are required. The goals of finding ways to include a broader and more diverse set of students, making science accessible across gender, cultural, economic and ethnic boundaries, and keeping students interested in STEM careers has foregrounded our understanding that thinking and working in innovative ways is necessary. Interdisciplinarity is one way to increase the available options and generate creative ideas.
2. *The importance of assessment*: Our educational programs are based on the assumption that teaching is an activity that can be empirically described, investigated, and explicated. This understanding of STEM education as a research topic that requires the systematic collection of evidence about student learning outcomes has meant that experts in assessment are integrated within our educational programs. Applied linguistics offers a range of methodological options useful for exploring student understanding in this setting.
3. *Engaging in issue-driven projects and being open to interdisciplinary methodological answers:* Our collaboration was greatly facilitated by the fact that the initial approach to the development of innovative science education programs consisted of open discussion between scientists and the funding agency. The openness of the discussion and the focus on issues and questions allowed the HHMI professors the freedom to consider new ways of answering these questions. This culture of freedom to explore creatively and focus on core issues facilitated interdisciplinarity and allowed new methodologies to be applied (as in the integration of applied linguistics within STEM).

4. *Being open to shared learning and the distribution of knowledge:* Working in interdisciplinary teams requires a willingness to learn from others and not be disturbed by the presence of expertise beyond one's own. Not knowing can be uncomfortable but also offers the possibility for personal growth. Applied linguists cannot be expected to know areas of STEM before working in these areas; STEM professionals cannot be expected to fully understand issues of applied linguistics. Both parties need to agree upon their shared mission and the core questions and issues to be addressed. The result will be an innovative integration and application of knowledge that provides novel answers.

We hope that this chapter, based on an existing collaboration, offers insight into ways in which applied linguists can work with STEM professionals. Ultimately our aim is to facilitate innovation in STEM education and, based on our shared experience, we believe that the enhanced use of applied linguistics expertise is one avenue for achieving this aim.

References

American Association of Medical Colleges – Howard Hughes Medical Institute. (2009). *Scientific foundations for future physicians.* ⟨http://www.hhmi.org/grants/pdf/08-209_AAMC-HHMI_report.pdf⟩

American Association for the Advancement of Science. (2010). *Vision and change: A call to action.* Washington, DC: Author. ⟨http://visionandchange.org/files/2010/03/VC_report.pdf⟩.

Anderson, W.A., Banerjee, U., Drennan, C.L., Elgin, S.C. R., et al. (2011). Changing the culture of science education at research universities. *Science, 334,* 152–153.

Clark, I.E., Romero- Calderón, R., Olson, J.M., Jaworski, L., & Lopatto, D. (2009). "Deconstructing" scientific research: A practical and scalable pedagogical tool to provide evidence based science instruction. *Public Library of Science: Biology, 7*(12), e1000264. doi:10.1371/journal.pbio.1000264.

Hanauer, D.I., & Bauerle, C. (2012). Facilitating innovation in science education through assessment reform. *Liberal Education, 98*(3). Retrieved from ⟨http://www.aacu.org/liberaleducation/le-su12/hanauer_bauerle.cfm⟩.

Hanauer, D.I., Hatfull, G.F., & Jacobs-Sera, D. (2009). *Active assessment: Assessing scientific inquiry.* New York, NY: Springer.

Hanauer, D.I., Jacobs-Sera, D., Pedulla, M., Cresawn, S.G., Hendrix, R.W., & Hatfull, G.F. (2006). Teaching scientific inquiry. *Science, 314*(5807), 1880–1881.

Handelsman, J., Miller, S., & Pfund, C. (2007). *Scientific teaching.* New York, NY: W. H. Freeman.

National Research Council. (2003). *BIO2010: Transforming undergraduate education for future research biologists.* Washington, DC: The National Academies Press.

National Research Council. (2009). *A new biology for the 21st century.* Washington, DC: The National Academies Press.

President's Council of Advisors on Science and Technology. (2012). *Prepare and inspire: K-12 education in science, technology, engineering and math (STEM) education for America's future.* Washington, DC: Author. ⟨http://www.whitehouse.gov/sites/default/files/microsites/ostp/pcast-engage-to-excel-final_2-25-12.pdf⟩

Index

2503240365